COMPUTER AIDED ASSESSMENT OF MATHEMATICS

Computer Aided Assessment of Mathematics

CHRIS SANGWIN

School of Mathematics, University of Birmingham

OXFORD
UNIVERSITY PRESS

Great Clarendon Street, Oxford, OX2 6DP,
United Kingdom

Oxford University Press is a department of the University of Oxford.
It furthers the University's objective of excellence in research, scholarship,
and education by publishing worldwide. Oxford is a registered trade mark of
Oxford University Press in the UK and in certain other countries

© Chris Sangwin 2013

The moral rights of the author have been asserted

First Edition published in 2013

Impression: 1

All rights reserved. No part of this publication may be reproduced, stored in
a retrieval system, or transmitted, in any form or by any means, without the
prior permission in writing of Oxford University Press, or as expressly permitted
by law, by licence or under terms agreed with the appropriate reprographics
rights organization. Enquiries concerning reproduction outside the scope of the
above should be sent to the Rights Department, Oxford University Press, at the
address above

You must not circulate this work in any other form
and you must impose this same condition on any acquirer

British Library Cataloguing in Publication Data

Data available

ISBN 978-0-19-966035-3

Printed and bound by
CPI Group (UK) Ltd, Croydon, CR0 4YY

Links to third party websites are provided by Oxford in good faith and
for information only. Oxford disclaims any responsibility for the materials
contained in any third party website referenced in this work.

CONTENTS

List of Figures — viii

1 **Introduction** — 1
 1.1 Multiple-choice questions — 2
 1.2 Assessment criteria — 4
 1.3 Chapters — 7
 1.4 Acknowledgements — 8

2 **An assessment vignette** — 9
 2.1 The student's perspective — 9
 2.2 Assessing answers to simple questions — 14
 2.3 Further integrals — 16
 2.4 Discussion — 18

3 **Learning and assessing mathematics** — 19
 3.1 The nature of mathematics — 19
 3.2 Terms used in assessment — 21
 3.3 Purposes of assessment — 22
 3.4 Learning — 23
 3.5 Principles and tensions of assessment design — 25
 3.6 Learning cycles and feedback — 33
 3.7 Conclusion — 35

4 **Mathematical question spaces** — 37
 4.1 Why randomly generate questions? — 38
 4.2 Randomly generating an individual question — 39
 4.3 Linking mathematical questions — 42
 4.4 Building up conceptions — 44
 4.5 Types of mathematics question — 46
 4.6 Embedding CAA into general teaching — 49
 4.7 Conclusion — 51

5 **Notation and syntax** — 53
 5.1 An episode in the history of mathematical notation — 54
 5.2 The importance of notational conventions — 56
 5.3 Ambiguities and inconsistencies in notation — 60
 5.4 Notation and machines: syntax — 61
 5.5 Other issues — 65
 5.6 The use of the AiM system by students — 66

	5.7 Proof and arguments	67
	5.8 Equation editors	68
	5.9 Dynamic interactions	70
	5.10 Conclusion	71
6	**Computer algebra systems for CAA**	**73**
	6.1 The prototype test: equivalence	75
	6.2 A comparison of mainstream CAS	76
	6.3 The representation of expressions by CAS	78
	6.4 Existence of mathematical objects	82
	6.5 'Simplify' is an ambiguous instruction	86
	6.6 Equality, equivalence, and sameness	88
	6.7 Forms of elementary mathematical expression	91
	6.8 Equations, inequalities, and systems of equations	94
	6.9 Other mathematical properties we might seek to establish	96
	6.10 Buggy rules	97
	6.11 Generating outcomes useful for CAA	99
	6.12 Side conditions and logic	100
	6.13 Conclusion	101
7	**The STACK CAA system**	**102**
	7.1 Background: the AiM CAA system	102
	7.2 Design goals for STACK	103
	7.3 STACK questions	106
	7.4 The design of STACK's multi-part tasks	107
	7.5 Interaction elements	111
	7.6 Assessment	112
	7.7 Quality control and exchange of questions	113
	7.8 Extensions and development of the STACK system by Aalto	114
	7.9 Usage by Aalto	117
	7.10 Student focus group	121
	7.11 Conclusion	125
8	**Software case studies**	**127**
	8.1 Some early history	127
	8.2 CALM	129
	8.3 Pass-IT	132
	8.4 OpenMark	138
	8.5 DIAGNOSYS	140
	8.6 Cognitive tutors	146
	8.7 Khan Academy	147
	8.8 Mathwise	148
	8.9 WeBWorK	150
	8.10 MathXpert	154
	8.11 Algebra tutors: Aplusix and T-algebra	157
	8.12 Conclusion	160

9	**The future**	**162**
	9.1 Encoding a complete mathematical argument	162
	9.2 Assessment of proof	166
	9.3 Semi-automatic marking	169
	9.4 Standards and interoperability	170
	9.5 Conclusion	172
Bibliography		173
Index		183

LIST OF FIGURES

1.1	Incorrect long multiplication.	2
1.2	A sample question from the STACK CAA system.	4
2.1	A simple STACK CAA question.	10
2.2	Validation of a simple answer.	11
2.3	Feedback based on properties.	12
2.4	Invalidity for mathematical reasons.	12
2.5	Incomplete answers: an opportunity for partial credit.	13
2.6	A correct answer.	13
2.7	A general test for indefinite integration problems.	17
4.1	A sequence of simple questions	43
4.2	Dynamic mathematics with GeoGebra.	50
5.1	The DragMath equation editor.	68
5.2	Dynamic interactions.	70
5.3	Sketching the derivative in WeBWork.	71
6.1	Mathematical objects and their representation.	83
6.2	Feedback about the form of an answer.	92
6.3	An answer which is a system of equations.	95
7.1	An AiM question, as seen by a student.	103
7.2	Many parts, one assessment algorithm.	108
7.3	Many parts, independent algorithms.	108
7.4	Many parts, many algorithms.	109
7.5	An example of display from the Aalto STACK system.	115
7.6	Muti-part question with randomly generated diagrams.	116
7.7	A question including a GeoGebra applet.	117
7.8	The grading system on the course discrete mathematics	119
7.9	Student scores from examinations and exercises	120
8.1	First phase of the CALM CAA system.	130
8.2	A Pass-IT question, before steps are revealed to the student.	133
8.3	A Pass-IT question, after steps have been revealed to the student.	133
8.4	A Pass-IT question in which part of the answer is a graph.	134
8.5	A SCHOLAR question in which part of the answer is a pie-chart.	135
8.6	Progressive feedback from OpenMark.	139
8.7	Mathinput panel from the DIAGNOSYS system.	141
8.8	The DIAGNOSYS system.	141

8.9	DIAGNOSYS algebra skills.	143
8.10	DIAGNOSYS skills.	144
8.11	Skill lattice for subtraction.	145
8.12	An example of the Cognitive Tutor.	147
8.13	Mathwise vectors module: the divergence quiz.	150
8.14	The user interface in the MathXpert system.	155
8.15	Tracking side conditions in the MathXpert system.	156
8.16	The Aplusix system.	158
8.17	Backtracking in Aplusix.	159
9.1	Proof in the EASy system.	167

1

Introduction

My friend Herschel, calling upon me, brought with him calculations of the computers, and commenced the tedious process of verification. After a time many discrepancies occurred, and at one point these discordances were so numerous that I exclaimed 'I wish to God these calculations had been executed by steam!' (Charles Babbage, cited in Buxton and Hyman, 1987)

This book examines automatic computer aided assessment (CAA) of mathematics. It is written principally for colleagues who teach mathematics to students in the later years of school and post-compulsory mathematics, including in universities. This book has two purposes:

1. To explain how CAA is currently being applied to assess mathematics.
2. By considering CAA, to discuss assessment of mathematics more generally.

The impetus for the Industrial Revolution was a desire to automate routine processes. For example, Charles Babbage was motivated to relieve human 'computers' from the effort of creating logarithm tables by hand; the result was his mechanical computers. Assessing students' work can be similarly repetitive, technical work, but one requiring expertise. The teacher has to make many judgements rapidly, and with the large class sizes typical in universities there are real limits on the quantity of work each student can expect the teacher to assess. While it takes one hour to lecture to 500 students, a working week gives each student less than five minutes of individual attention from a lecturer. One solution is to delegate assessment to teams, but this is problematic. Each team member is a human with their own ideas of what is acceptable: if the marks are used to record the achievement of the student, i.e. summative assessment, then consistency needs to be ensured. Moreover, large quantities of such repetitive work is arguably not the best use of a mathematical expert's time.

Between 2000 and 2011 at the University of Birmingham, the UK MSOR Network[1] organized a two-day residential workshop for staff new to teaching mathematics in higher

---

1. The Maths, Stats & OR Network supported lecturers in mathematics, statistics, and operational research in promoting, disseminating, and developing good practice in learning and teaching across the United Kingdom.

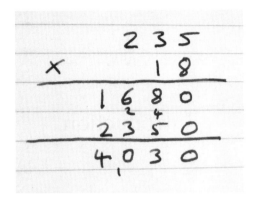

Figure 1.1: Incorrect long multiplication.

education, attracting around 30 colleagues annually. The participants were mostly new staff, but with some experienced colleagues new to teaching in the UK. Assessment was a key concern at this event and we began addressing it by asking colleagues to assign a mark from 0 to 10 for the work shown in Figure 1.1. You might like to do the same now.

There is always a variety in marks, from 0 (*'the answer is wrong'*) up to 9 (*'you can see they basically have the correct idea and have made only one silly slip'*). This surprises colleagues, many of whom have reached their conclusions after careful thought, and by drawing on sound principles. The context matters: colleagues often revise their mark when told, 'now this student nurse is calculating a drug dose'. There is some debate about what we mean by good and poor quality and how we distinguish it. There is even more scope to discuss what feedback will better support students.

Computer aided assessment is as old as computers themselves, and we shall examine the history of the field as relevant to mathematics. Furthermore, many people are working in this field: computer aided assessment is growing into an industry. Learned societies are taking an interest in this area; e.g. see Kehoe (2010). Both general systems and those targeted at specific subjects are being developed, with the inevitable duplication and reinvention. It is timely to record the current state of this art.

> The issue for e-assessment is not if it will happen, but rather, what, when, and how it will happen. E-assessment is a stimulus for rethinking the whole curriculum, as well as all current assessment systems. (Ridgeway *et al.*, 2004)

1.1 Multiple-choice questions

Automatic assessment is commonly associated with multiple-choice questions (MCQ). Indeed, many existing generic CAA systems provide only types of interaction in which potential answers provided by the teacher are selected by students. The most common example is a multiple-choice question in which a student selects a single response as their answer.

> If popularity is the sole criterion, no other test format for measuring ability and achievement can match that of the multiple-choice format. Its frequency of use in educational, industrial, and governmental institutions undoubtedly surpasses that of all other objective test types, and it is used more often than tests using essay items. (Hassmén and Hunt, 1994)

There are many related 'types'; for example, a multiple-response question (MRQ) in which a student chooses as many or as few of the available options as they need. Sometimes students indicate their confidence as well as their answer.

All these types of question are referred to as *objective questions*, and tests using them as *objective tests*. This is because the outcome is independent of any bias by the assessor or the examiner's own beliefs. The term 'objective test' is sometimes synonymous with multiple-choice questions; however, the automatic computer aided assessment which is the focus of this book is also objective in this sense.

A well-constructed MCQ presents a list of plausible *distracters*, which ideally will be constructed from a knowledge and understanding of common student errors. There are examples of situations in which such questions can be effective, and we shall examine some of these in more detail. In general it is very difficult to write effective MCQ items. In many situations the teacher is essentially forced to 'give the game away' by presenting choices up front. The student then has only to select or verify rather than create. In mathematics the purpose of many questions is grotesquely distorted by using a MCQ, since the difficulty of a reversible process is markedly altered in different directions. For example, solving an equation from scratch is significantly different from checking whether each potential response is indeed a solution. Expansion versus factorization of algebraic expressions, or integration versus differentiation, are further examples. What many of these examples have in common is the difficulty of an *inverse operation* relative to the direct operation. The strategic student does not answer the question as set, but checks each answer in reverse. Indeed, it might be argued that it is not just the strategic but the *sensible* student, with an understanding of the relative difficulties of these processes, who takes this approach. This distortion subverts the intention of the teacher in setting the question, so that we are not assessing the skill we wish to assess. Hence the question is *invalid*, a term we define in Section 3.3.

There are other problems with the MCQ format. It is possible that the student will remember the distracter and not the correct answer, thereby the testing process could well lead to incorrect learning. Authors such as Hoffmann (1962) go as far as saying that MCQ tests *'favour the nimble-witted, quick-reading candidates who form fast superficial judgements'* and *'penalize the student who has depth, subtlety and critical acumen'*. Further, it is claimed by Hassmén and Hunt (1994), Leder *et al.* (1999) and others, that the MCQ format itself has inherent gender bias.

To avoid these problems with the MCQ and similar question types *we give a strong preference to a system which evaluates answers provided by students* and which consist of their own mathematical expressions.

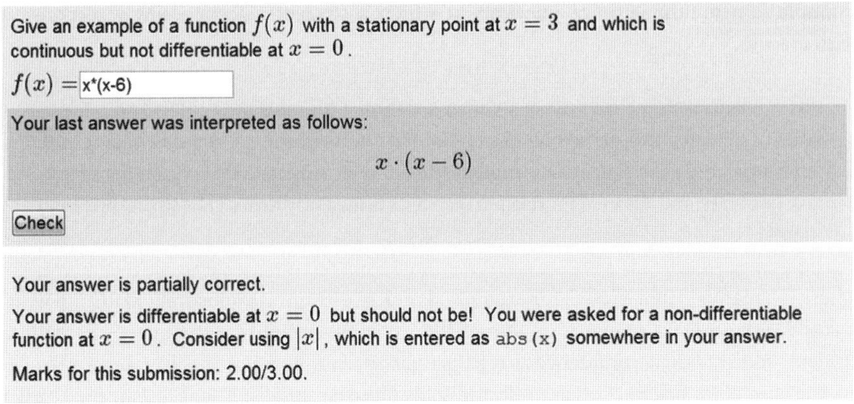

> Give an example of a function $f(x)$ with a stationary point at $x = 3$ and which is continuous but not differentiable at $x = 0$.
>
> $f(x) =$ `x*(x-6)`
>
> Your last answer was interpreted as follows:
>
> $$x \cdot (x - 6)$$
>
> [Check]
>
> Your answer is partially correct.
>
> Your answer is differentiable at $x = 0$ but should not be! You were asked for a non-differentiable function at $x = 0$. Consider using $|x|$, which is entered as `abs(x)` somewhere in your answer.
>
> Marks for this submission: 2.00/3.00.

Figure 1.2: A sample question from the STACK CAA system.

In this book we shall address the question of how we might automate the process of assessing a *mathematical expression provided by the student*. This will be the focus of our work. When a student enters their answer, e.g. an equation, they are not choosing an expression provided by someone else. Figure 1.2 is one example of such a question, together with a student's response and the associated feedback. In this case the assessment system is underpinned by a computer algebra system (CAS). This provides the teacher with a set of tools with which to manipulate the answer of the student and so establish relevant properties in an objective way. In order to do this we now have two further issues to address:

- How does the student enter their answer into a machine?
- How does the machine establish mathematical properties of such answers?

The first of these is addressed in Chapter 5 and the second is addressed in Chapter 6.

1.2 Assessment criteria

Experience has also shown that attempting to automate assessment raises some very interesting issues. For example, in order to automate the assessment of even a single final answer, such as that of Figure 1.2, it is necessary to codify the criteria that are acceptable. This is not so simple. Even writing pseudocode on paper to capture precisely the desired criteria forces the teacher to think very carefully indeed. Often there are a number of criteria: e.g. 'the student's answer is algebraically equivalent to mine', or 'the student's answer is simplified to a specified form'. How strict are you going to be in requiring all these? In Figure 1.2 three independent criteria are checked, and outcomes assigned accordingly. What feedback would students find helpful when only some are satisfied? With CAA all these decisions need to be made carefully and explicitly *in advance*. When writing the mark scheme of an examination paper we also have to decide the criteria, but not the feedback or how to deal with unexpected answers.

It may well be, of course, that the teacher *really* does want a specific expression, and that only this will be satisfactory. This is often a reasonable view for a teacher to adopt. After all, *canonical forms*, e.g. gathering like terms, ordering and simplifying coefficients of polynomials, are used precisely as conventions so that humans can easily compare mathematical expressions. Habitual use of these is part of mathematical culture, and therefore we naturally expect students to use them. Even in this case, the desire to award partial credit or to provide useful formative feedback forces a teacher to consider what properties contribute to this overall correctness. Some properties form the goals of the question, while others are underlying conventions.

One might know, in advance, that certain misconceptions or technical errors lead to particular answers. For the purposes of formative assessment, a term defined in Section 3.3, feedback could be given which suggests this to the student. They may be encouraged to attempt the question again. Identifying these, through professional experience, or collaboration with educational researchers, remains a challenge. Writing feedback which gets to the heart of the matter and which encourages people to think, without telling them the answer, is difficult.

When the only available evidence is the final answer it is not clear how to identify which method a student is using, or to check that this method is appropriate. We shall examine how current CAA deals with 'steps in working' and in identifying which method has been used, since it is a rather important issue. This area is where automatic assessment merges with 'intelligent tutoring'.

In generating random versions of individual questions, the teacher is forced to consider very carefully the purpose of the question within the scheme of work. What can be changed and what should remain the same to preserve this purpose? If the questions are structured to include particular cases then enthusiastic, or perhaps careless, randomization destroys this structure.

The student must enter an answer into a machine, as in Figure 1.2. In some cases this will be a multiple-choice response, or similar, which is technically rather trivial to implement and relatively easy for students to use. Many assessment systems accept answers that are mathematical expressions. That is to say, objects such as polynomials, equations, sets, lists, inequalities, and matrices. Now, the student is forced to be as explicit as the teacher in what they mean when entering their answer. This is not confined to the mismatch between traditional written notation $|x|$ and computer syntax `abs(x)` shown in Figure 1.2. Notation or syntax might appear trivial, but a close examination of traditional mathematical notation reveals some ambiguities and inconsistencies. There are also cultural differences. We shall examine this problem in some detail, but currently this is the most significant barrier to CAA use and even to mathematical human–machine interaction in general.

CAA also provides opportunities not present in traditional paper environments.

1. Questions can be randomly generated for each student. Originally intended to reduce impersonation of one student by another and plagiarism, this also provides possible additional practice of similarly structured problems and groupwork, with each student able to provide evidence of their own individual work. See Chapter 4.

2. Tests can be *adaptive*, where items (i.e. questions) are not fixed in a linear fashion. The next item depends on the student's response to the current one, or their overall profile of achievement.

3. Dynamic interactions can be made available. For example, dragging geometric configurations allow quite different experiences from a static diagram on paper. These configurations might constitute part of an answer.

None of these are possible in a fixed, traditional paper problem sheet when used with very large numbers of students.

The constraints of paper-based formats have affected what we do and why. It is natural for busy teachers to set only those questions which can be marked in a straightforward way. There are other questions, particularly those with non-unique correct answers, or where establishing the required properties requires the markers themselves to undertake a significant computation. It is simply not sensible for a person to set such questions to large groups of students when marking by hand. Yet such questions have their place and value in provoking thought and learning. Figure 1.2 is one example, and others will be provided in due course.

In a traditional paper environment, *follow-through marking* is sometimes used to take an incorrect answer and use this in subsequent working. This is an artifact of the paper-based format: we may want to reward what the student has achieved, and a single minor slip early in a long question might obscure a basically sound understanding. We could certainly, from a purely technical level, automate such follow-through marking in many cases. A very basic example is shown in Figure 7.4. However, during CAA the feedback is available almost immediately. There is no long delay while the teacher marks all the work and returns the paper to the students. Furthermore, a CAS can be programmed with the required 'steps' in the teacher's method, and where a question is broken down into stages, the student's answer to each used to generate the next automatically. Did we ever want follow-through marking anyway? Accuracy is an important goal. If the student knows their answer is wrong, should not they identify and fix the errors themselves and try again? We hope to examine some of the messages our assessment strategies send to students, and the ways the students change their behaviour as a result. We can perhaps harness some of these to achieve better outcomes for our students.

We do not propose CAA as a panacea. Many valuable mathematical tasks cannot be marked automatically, particularly proofs such as those which might be encountered in a real analysis or geometry course. Assessment of extended projects, history essays, and similar work will probably always require an expert. To concede that a tool does not do everything is quite different from saying it does nothing useful. We hope this book explains what is possible, in a pragmatic way, but also looks further and considers deeper questions of assessment. Do we simply want to replicate in CAA the traditional approaches used on paper, or should we reassess what is the most valuable use of these new tools for our students?

There are a number of related topics which we shall also consider in passing, which essentially require students to use technology other than pencil and paper. These include communication technology, e.g. 'e-learning', 'learning environments' and 'micro-worlds'.

This book is not about the rather thorny issue of students using technology to answer questions, e.g. calculators, and computer algebra systems. This issue is orthogonal to our concerns, and has been addressed in many other places. For one view see Buchberger (1990).

1.3 Chapters

Each chapter in this book is described below, and some of the questions these chapters will address are raised. Throughout, we will provide case studies to illustrate the practicality of the claims made and draw on research studies in mathematics education and computer aided assessment.

2 *An assessment vignette*
By examining a standard, procedural, single-step question, such as 'Find $\int 3x^5 dx$', we introduce and motivate further discussion of computer aided assessment.

3 *Learning and assessing mathematics*
In this chapter we attempt to answer the following questions. What does it mean to 'learn mathematics'? How do we know when someone has 'learned'? What is the role of practice in learning? What are we practising and why? Why do we assess? What role(s) does it play? How do we recognize what we have really assessed? How does the assessment process itself affect students' activities?

4 *Mathematical question spaces*
It is natural to split an individual mathematical assignment into tasks. We look at structure within sequences of such tasks. We also consider various forms of individual tasks used in mathematics teaching. In CAA these tasks can often be generated randomly, but in doing so we need to preserve their essential character.

5 *Notation and syntax*
Mathematical notation has a long and interesting history. It also has a profound influence on the mind. For mathematical CAA, students must enter their answer into the machine. In this chapter we examine:
- Mathematical notation and its meaning.
- Syntax for typing expressions into a machine.
- Other types of interaction.

6 *CAS for CAA*
Computer algebra systems (CAS) automate computation and so are central to automatic assessment. In automating mathematical computation we reveal very interesting issues in elementary mathematics. This chapter considers the design and use of CAS for CAA.

7 *The STACK CAA system*
We bring together the theoretical work of the previous chapters and apply this to practical assessment. In particular, we look in detail at the STACK system as a case study in CAA. This chapter uses specific examples without the detail that a

software manual might provide. We also provide a detailed case study of the use of the STACK CAA system.

8. *Examples of CAA systems*
   We compare STACK with the work of others in mathematical CAA. In particular we consider how question authors have implemented the automation of assessment of steps in working and look at examples of adaptive testing.

9. *The future: assessing steps in free text, proofs in geometry, real analysis*
   Work in this field is progressing rapidly. We consider some pilot studies in assessing free response steps in calculations, and also proofs in real analysis and geometry.

Computer aided assessment is here, and here to stay. The number of students and their teachers using online assessments is increasing rapidly. We hope this book provides teachers with interesting background to this topic and examples of current practice. In preparing this book, in particular in undertaking the background research for Chapters 8 and 9, it has become clear quite how many different independent groups are working in this area. Ultimately the most important issue is not the features provided by the tool, but how and why it is used. This responsibility remains with teachers. The purpose of this book is to help them make discerning and informed choices in this area.

1.4 Acknowledgements

The work underlying this book, particularly the software development of STACK and the practical implementation during teaching, has been undertaken in collaboration with a number of colleagues. Dirk Hermans in the School of Mathematics at the University of Birmingham introduced the AiM CAA system (see Section 7.1) from which STACK grew. At the University of Birmingham, code for CAA systems have been developed with help from Laura Naismith, Jonathan Hart, and Simon Hammond. The core code for STACK version 3 was written by the author, and tightly intergraded into the Moodle quiz by Tim Hunt of the UK Open University. The support of Phil Butcher of the Open University is very gratefully acknowledged. Other significant code contributions have been written by Finnish colleagues Matti Harjula and Matti Pauna. There have been many other contributors, including those who have provided translations. STACK also relies on numerous libraries of code, in particular on Maxima, which have been developed and generously released by a very large number of people. The author would also like to thank colleagues at Aalto, Helsinki, for generously providing their data on the use of STACK and facilitating the focus group reported in Section 7.10. The contribution of the four students who participated in the focus group is particularly acknowledged.

Chapter 8 contains many figures illustrating existing CAA systems, and the author is grateful to colleagues who supplied them.

2

An assessment vignette

This short chapter provides a vignette of automatic computer aided assessment of mathematics. Initially we consider the student's point of view before discussing the assessment criteria used by the teacher. The example chosen is taken from the STACK CAA system designed and implemented by the author, and discussed in more detail in Chapter 7. The purpose of this vignette is to provide a concrete example of current practice from which we can illustrate the theoretical and practical issues which form the topics of subsequent chapters.

Imagine a methods-based mathematics course in which a group of students is learning the techniques of symbolic integration. The students have seen differentiation of algebraic expressions in a single variable, and they are practised in finding the derivatives of simple polynomials. They also know various general rules for differentiation such as the chain rule.

Students have been taught 'integration as the reverse of differentiation' as a procedure by which to actually calculate an integral symbolically. Their task is to practice this. A formative computer aided assessment quiz is used, consisting of items such as the following:

$$\text{Find} \int 3x^5 dx. \tag{2.1}$$

Our task, for the purposes of illustration, is to automate the assessment of answers to this task. This example is deliberately a very simple elementary task, and is by no means the limit of what is possible with CAA. We note that the extent to which students should be able to calculate such a symbolic integral is also not the focus of this chapter.

2.1 The student's perspective

In order to actually answer the questions the student has to *authenticate* themselves. Normally, students will be provided with a username and password for a networked application or access to a website, or simply make use of open access materials which do not track a user's actions. There are many online systems for managing courses. They define roles, e.g. students and teachers, and hence control access to materials and activities. They also record users' actions. Moodle (http://www.moodle.org) is an online content management system. STACK is a mathematical question type for Moodle's quiz. Hence, to interact with STACK questions a student enrols on a course which contains a quiz. For the purpose of illustration we shall assume that this quiz contains question (2.1) and that the student has navigated to the web page shown in Figure 2.1. The page contains the question, the context in which the question is set, i.e. the subject, and the quiz name (Vignette quiz). Notice the page contains the student's name. It is very useful in high-stakes invigilated computer assessments for the invigilator to immediately see, on the screen on which the student is working, the name of the student who will be credited with this work. Since there is no handwriting evidence, detecting impersonation can be a problem.

In this situation the quiz consists of a fixed list of questions, and the student may move freely between them. In other circumstances the questions might be selected randomly from a pool, or during *adaptive testing* the next question is determined from the outcomes to previous attempts by building a *user profile*. Here the teacher has chosen to give students the option to repeatedly attempt the quiz and also make multiple attempts at individual questions within this. In both cases the student can finish a session and return later. Following the 'Finish attempt . . . ' link the student could ask the system to create a completely fresh quiz with potentially new random versions.

The question in Figure 2.1 contains one *interaction element*, in the form of a text box into which the student should enter their answer, e.g. 3x^6. The *client-server* model on which web-browsing technology is based forces a particular kind of interaction in which a user presses the 'Check' button to submit a page. The next page is based on the values of such interaction elements. If the student's answer is $3x^6$ then the response is shown in

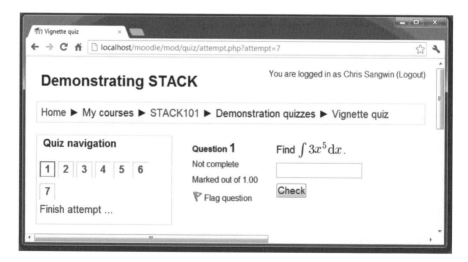

Figure 2.1: A simple STACK CAA question.

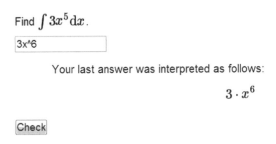

Figure 2.2: Validation of a simple answer.

Figure 2.2. Note that the student has not typed 3*x^6 but has used *implicit multiplication* and typed 3x^6 instead. The system has echoed their one dimensional input in traditional two-dimensional mathematical notation and has indicated the implied multiplication with a centre dot: $3 \cdot x^6$. The system has also judged this input as *valid*. The difference between *validity* and *correctness* will be discussed in detail later. An answer must be valid before the mathematical properties can be established. Incorrect use of syntax, e.g. mismatched parentheses, would render an expression invalid.

The student may change their answer at this stage and submit the page again. Once the student is satisfied that the system has interpreted their answer correctly, and that it is valid, a further 'Check' request will assess the answer. The feedback shown in Figure 2.3 contains detailed formative feedback which is based on the mathematical properties of the expression typed in by the student. In particular, the CAS has taken the expression entered by the student, differentiated this, and asked the student to compare this with the integrand. The student is now able to consider this feedback, and in the light of this, try again. The crucial point is the mathematical calculation using the student's answer and the feedback which incorporates this.

Let us assume that the next attempt the student tries is $\frac{3}{6}x^6$, which is entered as 3/6*x^6. The result is shown in Figure 2.4. Notice that this result is considered invalid, not because of a syntax error but for a mathematical reason. Here the goal of the question is not to cancel fractions, but to do so when giving an answer remains an important underlying convention. The teacher has chosen to render certain expressions invalid, rather than wrong, to reinforce this particular pedagogic issue.

We shall assume that the student simplifies the expression correctly and has typed in $\frac{x^6}{2}$ and validated this. A further submission of the page then assesses this new answer, the results of which are shown in Figure 2.5. Notice that this answer is considered to be incorrect *only* because a constant of integration is missing. In this case the teacher has decided not to award *partial credit* for a missing constant of integration. A diligent student will read and act on the feedback. Availability of multiple tries provides every opportunity to correct such mistakes. Individual teachers need to take such decisions in the context of their particular teaching environments.

To finish the process the student simply needs to add a constant of integration, perhaps by entering an answer such as x^6/2+c. Submitting this page to validate the changed

Find $\int 3x^5 \, dx$.

Your last answer was interpreted as follows:

$$3 \cdot x^6$$

Check

Incorrect answer.

The derivative of your answer should be equal to the expression that you were asked to integrate, that was:

$$3 \cdot x^5$$

In fact, the derivative of your answer, with respect to x is:

$$18 \cdot x^5$$

so you must have done something wrong!
Marks for this submission: 0.00/1.00. This submission attracted a penalty of 0.10.

Figure 2.3: Feedback based on properties.

Find $\int 3x^5 \, dx$.

Your last answer was interpreted as follows:

$$\frac{3}{6} \cdot x^6$$

This answer is invalid.

Your answer contains fractions that are not written in lowest terms. Please cancel factors and try again.

Check

Figure 2.4: Invalidity for mathematical reasons.

answer, and again to assess it, results in the feedback shown in Figure 2.6. Since the teacher has decided to subtract 10% of the marks available for this question for each valid, but incorrect, different attempt the final mark for these attempts is 0.8. This mechanism provides an opportunity to answer a question again in the light of feedback. This 'penalty' scheme *rewards persistence and diligence* mitigated by the distinction between validity and correctness. It is, of course, unfortunate that small summative numerical marks seem to have such a profound motivating influence on students. Again, a choice has been made here

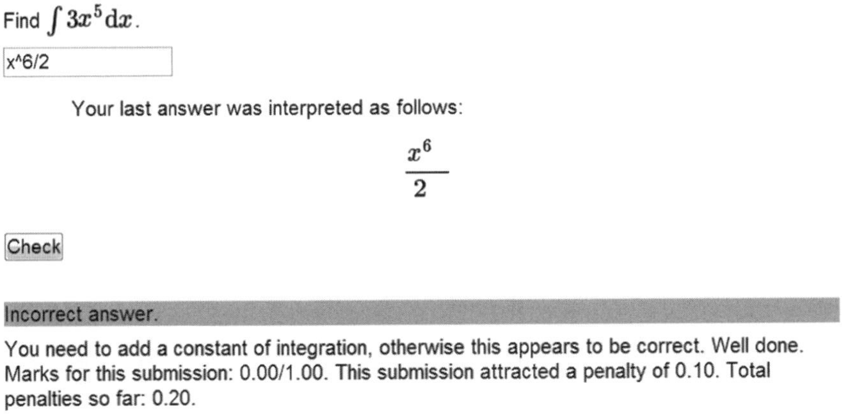

Find $\int 3x^5\,dx$.

`x^6/2`

Your last answer was interpreted as follows:

$$\frac{x^6}{2}$$

`Check`

Incorrect answer.
You need to add a constant of integration, otherwise this appears to be correct. Well done. Marks for this submission: 0.00/1.00. This submission attracted a penalty of 0.10. Total penalties so far: 0.20.

Figure 2.5: Incomplete answers: an opportunity for partial credit.

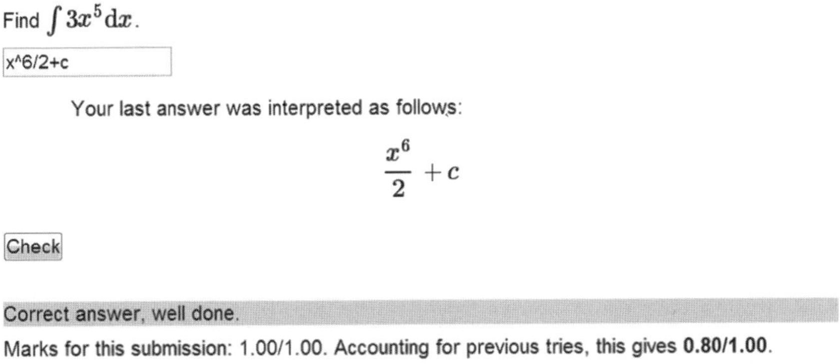

Find $\int 3x^5\,dx$.

`x^6/2+c`

Your last answer was interpreted as follows:

$$\frac{x^6}{2}+c$$

`Check`

Correct answer, well done.
Marks for this submission: 1.00/1.00. Accounting for previous tries, this gives 0.80/1.00.

Figure 2.6: A correct answer.

to provide both text-based feedback and a numerical score. A teacher could opt not to show numerical scores at all during formative work. There are potentially many other ways to generate numerical scores from a list of attempts.

Once all the questions are complete, students can see the teacher's answer and any detailed worked solution, known as *general feedback* in the Moodle quiz. Once the student has seen a worked solution they may make no further attempts at this question, but in this case may ask for a fresh randomly generated quiz.

The *feedback* provided to the student consists of four parts:

1 Their answer shown in traditional notation.
2 Information about the validity of the expression in the interaction element.

Following a repeat submission of a valid attempt:

3 Textual feedback about the attempt.
4 A numerical *score* for this attempt, and an overall score.

Internally there is also a 'note' which records the logical outcome regardless of any randomization in the question which is designed to facilitate later analysis by the teacher. There is also a worked solution, which is not 'feedback' in the truest sense of the term. At an early stage of learning a new topic, following such a worked solution through in detail can be particularly instructive. It is a way of getting to grips with a new method.

> Students spent far more time on the feedback than expected, resulting in them being able to do only two or three questions in a 50-minute test period rather than the five anticipated when writing the tests. (Greenhow and Gill, 2008)

Why is it important to provide such detailed feedback? Is it not sufficient to provide only a binary correct/incorrect outcome, an associated mark, and give a student a summary percentage at the end? The nature of effective feedback is discussed in Section 3.6. For now we note that the feedback shown in Figures 2.3, 2.4, and 2.5 is specific to the task and indicates how a student might improve their performance.

2.2 Assessing answers to simple questions

We turn our attention now to the problem of assessing answers to (2.1). It is reasonable to start the assessment process by looking at the student's *final answer*. What are the *mathematical properties* the answer should satisfy? Actually, there are quite a number of separate properties:

1 Is the answer an anti-derivative of the integrand?
2 Does the answer contain a constant of integration?
3 Is the answer 'fully simplified'?
4 Is the answer sufficiently general? I.e. is the answer complete?

The importance attached by the teacher to the answers to each of these questions will, naturally, vary according to the teaching context. Nevertheless, a teacher is likely to consider each in relation to a particular student's work. Essentially, the assessment process requires the teacher to make *many fine judgements rapidly*, and on the basis of these assign outcomes. Note that this process is quite different from asking '*does the answer look right?*', or equivalently, is the answer $\frac{1}{2}x^6 + c$? An essential difficulty in mathematical CAA lies in articulating those properties sought and robustly encoding tests to establish them. Next we draw a flow chart showing how the true/false answers to these assessment judgements result in outcomes. From this point on, the choices made are somewhat subjective. There is plenty of scope for discussion about what might be considered appropriate for groups of students in differing contexts.

The first test we consider, together with suggested feedback, is given below.

1.	Test:	Is the answer an anti-derivative of the integrand?
	if true:	
	if false:	The derivative of your answer should be equal to the expression that you were asked to integrate, which was: ex1. In fact, the derivative of your answer, with respect to v is ex2, so you must have done something wrong!

To undertake this test we use a computer algebra system to differentiate the student's answer with respect to v, the appropriate variable in the question, and compare this expression with the integrand. This solves the difficulty of removing a potential constant of integration, which might conceivably use any legitimate variable name. By 'compare' we actually mean establish *algebraic equivalence* of the two expressions. A last thing to note in this example is the proposed formative feedback. This feedback contains mathematical expressions relevant to the question, i.e. ex1 and v together with the results of performing a mathematical calculation with the CAS on the expression entered by the student, i.e. ex2. Obviously, the values of these expressions need to be substituted before the text is displayed to the student. Note, however, that this means the text must be generated in response to the actual answer given: it is genuine feedback to the student's answer.

Next we need to establish whether the student's answer has a constant of integration. If so, is this done in an appropriate way? Establishing this is technically complex, since we need to ensure the constant is general, but occurs in a linear fashion. Examples of cases are shown below.

2.	Test:	Does the answer contain a constant of integration?
	e.g. $\frac{1}{2}x^6 + c$	(Correct)
	e.g. $\frac{1}{2}x^6$	You need to add a constant of integration, otherwise this appears to be correct. Well done.
	e.g. $\frac{1}{2}x^6 + 1$	You need to add a constant of integration. This should be an arbitrary constant, not a number.
	e.g. $\frac{1}{2}x^6 + c^2$	The formal derivative of your answer does equal the expression that you were asked to integrate. However, you have a strange constant of integration. Please ask your teacher about this.

There are other issues, for example if a student's answer is $\ln(x)$ when the teacher expected $\ln(|x|) + c$ then the system will provide feedback: '*The formal derivative of your answer does equal the expression that you were asked to integrate. However, your answer differs from the correct answer in a significant way, that is to say not just, e.g., a constant of integration. Please ask your teacher about this.*' In Chapter 6 we shall discuss the limits of the extent to which properties can be established automatically.

So far we have concentrated only on establishing whether the student's answer is correct. If the student's answer is not an anti-derivative of the integrand we might consider the following:

3.	Test:	Has the student differentiated instead?
	if true:	It appears you have differentiated by mistake.
	if false:	

This is easily checked by differentiating the integrand and comparing it with the expression supplied by the student. Although a student has used the process of formal differentiation, they may still add a constant of integration anyway, which adds another level of complexity to implementing this test robustly.

There may be other mistakes, specific to the question, for which a teacher could check and provide tailored feedback in each case. The extent to which this is possible, or desirable, is very context dependent. Other likely mistakes for questions of the type $\int nx^m \, dx$ were identified by Sangwin (2003a) as, in order of frequency,

$$\frac{n}{m}x^{m+1}, \quad n(m+1)x^{m+1}, \quad \frac{n}{m+1}x^m.$$

In each case, formative feedback and partial credit could be specified. If none of the tests reveal what the student has done, we might supply some 'generic feedback' such as that of test 1. Perhaps we might choose to supply such feedback anyway. By providing some indication as to where he or she may have gone wrong, we provide a much greater incentive for the student to retry the question immediately, as opposed to simply stating the answer is incorrect. This is exactly the kind of task-based feedback suggested as most helpful by Kluger and DeNisi (1996).

2.3 Further integrals

The next example illustrates further the limitations of response matching at a string level, and provides additional justification for the mathematical sophistication enabled by systems such as STACK. Consider which of the following answers you would accept to the following question:

$$\text{Find } \int \frac{1}{3x} \, dx. \tag{2.2}$$

(i) $\frac{1}{3}\ln(x)$, (ii) $\frac{1}{3}\ln(|x|)$, (iii) $\frac{1}{3}\ln(x) + c$, (iv) $\ln\left(k\,|x|^{\frac{1}{3}}\right)$? The formal derivative of each of these equals the integrand, although arguably only with the last is the answer the correct anti-derivative, contains a constant of integration, and is 'fully simplified'. Furthermore, we might have two different answers[1] since $\ln(|x^{\frac{1}{3}}|) + c$ is arguably also acceptable in this case.

It is clear that assessing a syntactically valid, free-response, answer to question (2.1) becomes rather complex when the teacher tries to provide helpful, robust, formative feedback to the student. But formative feedback is cited as one of the benefits of CAA, and so this poses a problem to the individual teacher wishing to author their own questions.

One approach, e.g. taken by STACK and some other systems, is to provide tool kits of code which automates much of the process of providing feedback in commonly occurring questions. Indefinite integration, such as question (2.1), is a commonly occurring practice problem, for which CAA is particularly useful. Many of the tests discussed so far would be implemented regardless of the integrand, and to do so robustly it therefore makes sense to provide a test within the system which does this. Figure 2.7 shows the tree underlying such a test in the STACK CAA system, designed for indefinite integration. This tree also illustrates the order in which the tests are applied, and when it is felt appropriate to provide different feedback which might result from such tests. By providing such tests, the teacher is relieved of the need to consider these issues in detail for each question. Only those tests specific to the circumstances need be encoded.

TA is the teacher's answer.
SA is the student's answer.
v is the variable, with respect to which integration takes place.

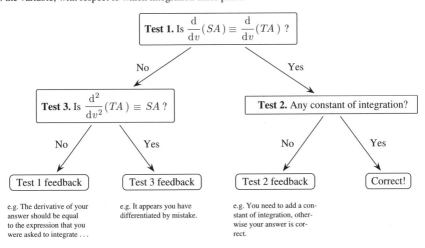

Figure 2.7: A general test for indefinite integration problems.

---

1. There is surprising variety of approaches to $\int 1/x dx$ within school textbooks. For an explanation of why $\int 1/x dx = \ln(|x|)$ see Urban et al., 2005, p. 656. Note that many computer algebra systems give $\int 1/x dx = \ln(x)$ instead.

2.4 Discussion

In working through this simple example we have illustrated a number of issues in contemporary CAA.

In practice it is common to randomly select a question from a pool, or to randomly generate parameters within a particular question (or both). In the example shown here we could randomly change the coefficient, power, and variable in an attempt to reduce the problems of plagiarism and impersonation. A detailed discussion of this issue is presented in Chapter 4.

Students must provide an answer which consists of a mathematical expression. The system then seeks to establish properties of this expression. In particular we have separated out the notions of validity from correctness. There are multiple mathematical properties which constitute *correctness* and a number of other properties which are underlying conventions. For example, writing rational numbers in lowest terms. The key difficulty for the teacher is in articulating these properties. Current mathematical CAA is sophisticated, and any teacher seeking to write their own questions needs to be prepared to invest a little time to understand the theoretical and practical background.

There is an underlying *interaction model* which is inherent and implicit. One layer is inherited from the client–server interactions used by forms on Internet pages in which a whole page is sent back to the server before the next page is generated in response. A desktop application might have very different kinds of interaction. For example, the equation might be built character by character as the student types each key. Another layer of the model concerns how multiple interactions with questions change the feedback. Various options exist to make individual questions available in a sequence. Here we have a fixed quiz, but in other situations we might have adaptive testing.

In Chapter 5 we shall consider mathematical notation and how a student types in their answer. In Chapter 6 we consider computer algebra to support establishing properties of an expression. Chapter 7 returns to STACK to consider the interaction model for individual questions, e.g. how we deal with multiple parts. Then we look at a case study for the actual use of STACK. We then compare the detailed case-study of STACK with many other systems in Chapter 8 before speculating on the direction of possible future developments in Chapter 9. Before we examine these topics it is necessary to discuss what it means to learn mathematics and how to assess learning.

3

Learning and assessing mathematics

> Propose to an Englishman any principle, or any instrument, however admirable, and you will observe that the whole effort of the English mind is directed to find a difficulty, a defect, or an impossibility in it. If you speak to him of a machine for peeling a potato, he will pronounce it impossible: if you peel a potato with it before his eyes, he will declare it useless, because it will not slice a pineapple. (Charles Babbage, cited in Babbage, 1948)

Computer aided assessment is a *tool* that can be used to *assess students' learning*. What do we mean by the words 'assess' and 'learning'? Before we can discuss the circumstances in which CAA might be used effectively we need to develop a vocabulary so that different aspects of this complex process can be distinguished. Only then will we know whether we are trying to deal with a 'potato' or a 'pineapple'.

It is difficult to make general valid statements of how to structure assessments at a very fine-grained level of detail. For example, if a course objective is modelling, then the assessments are likely to be very different from a course where the purpose is fluency with mental arithmetic. With this in mind, this chapter examines briefly the nature of mathematics, defines terms commonly used in the assessment literature, and then uses these to discuss principles and tensions which arise when we try to assess students' learning of mathematics.

3.1 The nature of mathematics

Mathematics is a systematic way of structuring thought and arguments, which is tied closely to a coherent body of associated knowledge. Mathematics has a very long continuous history: it arose to solve practical problems of tax, accountancy, and metrology in approximately 2500 BC, Robson (2008). Such problems motivated the need for the development from arithmetic to algebra, and these original problems often survive in a recognizable form today. Mathematics subsequently developed into a number of sub-disciplines, including pure mathematics, applied mathematics and statistics.

Pure mathematics is concerned with the study of mathematical structures for their own sake. This includes the fundamental assumptions, e.g. the definition of 'numbers', on which the discipline is built. It examines patterns the consequences of assumptions and connections between different areas. Such patterns are often beautiful, intriguing, and surprising. Pure mathematics also justifies when and how the standard algorithms can be used correctly. For example, how to solve equations of different types, how many solutions to expect, and the properties of such solutions. These algorithms form the core of the subject and are important cultural artifacts in their own right. Learning to use these algorithms correctly and fluently is key to progress in mathematics education.

Applied mathematics, including statistics, solves problems relevant to the real world. In simplified form, applied mathematics proceeds by iterating the following steps:

1. Information relevant to the problem must be identified and selected.
2. The relevant information is abstracted into a mathematical formulation (modelling).
3. Techniques, e.g. algebra or calculus, are applied to solve the mathematical formulation.
4. The mathematical results are interpreted in terms of the original problem, and predictions from the model are tested. The assumptions upon which relevance was decided are challenged.

Rather than attempt to distinguish between pure and applied mathematics we highlight the difference between deductive and empirical justification. Mathematicians justify their work deductively from stated hypotheses, whereas the experimental scientist looks for empirical evidence. Therefore, the mathematical *proof* of a result and *deductive justification* are central hallmarks of mathematics. The applied mathematician needs to use and acknowledge both: checking the validity of a modelling assumption against an observation of a physical system is an empirical process.

In both the pure and applied sphere, mathematicians often do not discover their results by working deductively. They use analogy, intuition, experiment, and their previous experience. Hence, the final product of mathematical activity, i.e. the theorem together with its proof, differs significantly from the process by which it is discovered. See, for example, Polya (1962) and Lakatos (1976).

Mathematics is unusual in the extent to which one topic builds directly upon another. Progress can be made without complete mastery of a prerequisite topic; indeed, it is argued by e.g. Hewitt (1996) that important forms of learning take place when a technique is used as part of a more complex process. Nevertheless, one of the hallmarks of mathematics is very highly structured knowledge.

Statistics is similar to applied mathematics in that doing statistics involves setting up a problem, planning the statistical methodology, actual data collection, presentation/analysis, and discussion/conclusions. In cases where the evidence is not *statistically significant* the cycle needs to be revisited. Notice, crucially, that the statistical design work precedes any data collection. See Marriott *et al.* (2009).

Historically, practical problems have provided much of the motivation for the development of the general methods of pure mathematics. For example, a physical system which is modelled by equations which cannot be solved motivates a new area of research. Once the original system has been solved, pure mathematical concerns generalize the techniques and seek connections to apparently unrelated areas.

3.2 Terms used in assessment

To *assess* is to establish some quantity, quality, value, or characteristic. For us then, *assessment* is the process by which a *teacher* forms a *judgement* about a *student* and on the basis of that judgement assigns *outcomes*, such as feedback and a numerical mark/score. We confine ourselves to judgements formed by considering their mathematical work. In this section we shall define some important terms, and in the next section we shall consider tensions which arise when we try to identify examples of 'good' assessments for students.

The noun *assessment* may be used to refer to the *instrument* itself, such as an examination paper, which is used within this process. We choose to avoid this potential ambiguity using assessment only as a process. An *item* is the smallest independent unit within this instrument; for example, an individual question, or the parts of a question about a single object. An *instrument* may consist of a number of *items*.

Because of the multi-faceted nature of teaching, the teacher may play many roles. Indeed, Morgan (1998) (see also Morgan *et al.* (2002)) identified the following:

- Examiner, using externally determined criteria.
- Examiner, setting and using their own criteria.
- Teacher–advocate, looking for opportunities to give credit to students.
- Teacher–adviser, suggesting ways of meeting the criteria.
- Teacher–pedagogue, suggesting ways students might improve their perceived levels of mathematical competence.
- Imaginary naive reader.
- Interested mathematician/reviewer.

In some circumstances the individual 'teacher' may be replaced by other agents, such as an examinations board, or parts of the process may be undertaken by different individuals. In computer aided assessment, part or all of this process is codified into a machine and the automation may also assume these roles. However, the human teacher always remains responsible for the decisions leading to the various outcomes.

We note that *peer-assessment* does fit with this framework, although here the roles played by individuals are more fluid. Rapid changes take place between those acting as teachers and those acting as students. That is to say, for the purposes of our discussions of assessment, whether someone is the teacher or student is defined by their current role in the process.

In *self-assessment* an individual acts as the student and then as teacher, although the criteria used may be obtained from elsewhere. There is a subtle difference between self-assessment and *reflection* on the outcomes of assessment.

Predominantly this judgement involves establishing whether a student's mathematics satisfies *properties* specified by the teacher. For us this is a key concern, and we shall examine in fine detail what properties a teacher may wish to establish and the extent to which CAA can support this automatically. For example, is the student's final answer the 'same' as that suggested by the teacher? If not, are there mistakes, or where does a student's work deviate from the teacher's 'model solution'?

The word *judgement* is certainly emotive: who likes to be judged? Nevertheless, it is an accurate technical use of the word. We acknowledge at this stage that such judgements may or may not be *objective* and in both cases they reflect the *values* of the individual teacher and may or may not be endorsed by a community of practice. While some authors, e.g. Morgan *et al.* (2002), use the word *evaluation* in place of judgement we reserve this for the assessment of the effectiveness of an assessment instrument.

It is helpful to separate the *form* of the instrument from the *purpose* of assessment. Examples of forms include written coursework, examinations, project work, *viva voce*, a show of hands in class, and so on. Our focus is automatic computer aided assessment.

3.3 Purposes of assessment

Assessment fulfills a number of purposes. Wiliam and Black (1996) suggest that the *purpose* of assessment is characterized by the use to which outcomes are put. Outcomes which are made available to the student are called *feedback*.

1. *Diagnostic assessment*
   This is designed to establish the extent of past learning. The feedback is often a detailed profile, explaining where a student should target effort, particularly revision, in preparation for future new work. For some, *diagnostic* is seen as a pejorative term, and the term *developing a model of the student* is used instead.

2. *Formative assessment*
   Formative assessment supports and informs students' learning. In this case, feedback could be *qualitative*; for example, written or oral comments. Such comments could be detailed and tailored to what the student has written, or brief indications of where students' written work departs from model solutions.

3. *Summative assessment*
   Summative assessment seeks to establish the achievement of the student. In mathematics, summative feedback is most often *quantitative*, either a mark or a percentage.

   The *stakes* of a summative assessment describe the future importance of this record of achievement; for example, in allowing progression to subsequent courses or institutions. High-stakes summative assessments are typically end-of-course examinations, which are timed and unseen.

   The *stake-holders* include students, teachers, and employers/institutions.

4. *Evaluative assessment*
   Evaluative assessment seeks to measure the effectiveness of the teaching or the assessment of students. The purpose of such assessments can be refined further

to *quality enhancement* or *quality audit* functions. Note that *evaluation* of teaching may include, but is not limited to, evaluative assessment involving students. See, e.g., Chatterji (2003).

Rarely does assessment of a single student take place. Normally an individual instrument will be used to assess a *cohort* of students. If the outcome of the assessment is based only on the response of each individual student matched to the teacher's criteria, then the outcome is said to be *criteria-referenced*. That is to say, the judgement formed by the teacher is based only on specific criteria. We again note that these criteria may be *objective* or *subjective*. However, if comparisons are made between responses of different students then the outcome is said to be *norm-referenced*. An extreme example would be an outcome consisting of a ranking of students in order of the relative merit of their work. A third option is *ipsative* assessment in which present performance is compare with the prior performance of the student being assessed.

Some authors, e.g. Chatterji (2003), stresses the importance of evaluating assessment to ensure validity, reliability, and practicality. Broadly speaking, an assessment is *valid* when it tests what it is supposed to. That is, it tests the *foci* of assessment, Niss (1993), sometimes called the *learning objectives*. It is *reliable* when it gives similar results under varying circumstances, and *practical* when it does not overstrain resources to the detriment of the assessment. It is very difficult to know with any confidence how an individual item of assessment will match up to evaluative criteria and so for high-stakes situations with large cohorts they should be determined in advance by undertaking trials.

These aspects of assessment have nothing specifically to do with mathematics. Similarly, CAA is a tool for implementing automatic assessment, and it has the potential to be used for a number of purposes. We are dealing with the assessment of mathematics, and so we now turn our attention to the nature of 'good' assessment, how we might recognize it, and ultimately how we might implement it automatically with computers.

3.4 Learning

What does it mean to learn mathematics? This is quite different from asking what it means to *teach*.

> I was at the mathematical school, where the master taught his pupils after a method scarce imaginable to us in Europe. The proposition, and demonstration, were fairly written on a thin wafer, with ink composed of a cephalic tincture.[1] This, the student was to swallow upon a fasting stomach, and for three days following, eat nothing but bread and water. As the wafer digested, the tincture mounted to his brain, bearing the proposition along with it. But the success has not hitherto been answerable, partly by some error in the quantum or composition, and partly by the perverseness of lads, to whom this bolus is so nauseous, that they generally steal aside, and discharge it upwards, before it can operate;

---

1. Curing or relieving disorders of the head.

neither have they been yet persuaded to use so long an abstinence, as the prescription requires. (Swift, 1726, Part III, Chapter 5)

While satire sometimes contains a grain of pragmatic truth, answering our question is very difficult.

> Once you enquire into the nature of learning, it is amazing how hard it is to pin down what you mean. In every generation, authors have tried because in order to develop effective teaching, and it is necessary to be able to identify and verify learning. (Mason and Johnston-Wilder, 2004)

They continue by discussing the answers other authors have given to this question, and consequently Mason and Johnston-Wilder (2004) is an invaluable resource. For pragmatic reasons we wish to keep this section short, and so begin our discussion with the following excerpt from the *Oxford English Dictionary*:

> **learn** *v.* **1.** *tr.* gain knowledge of or skill in by study, experience, or being taught. **2.** *tr.* acquire or develop a particular ability (*learn to swim*). **3.** *tr.* commit to memory (*will try to learn your names*). **4.** *intr.* be informed about. **5.** *tr.* become aware of by information or from observation.

This quote highlights a number of key features:

- Knowledge, and memory of knowledge.
- Skills, their acquisition and the process of acquisition.
- Awareness.

We also draw the distinction in learning mathematics between understanding concepts and developing proficiency in the application of mathematical techniques. The TELRI project Roach *et al.* (2001) differentiated between two forms of learning as follows. They defined *adoptive* learning as an essentially reproductive process requiring the application of well-understood knowledge in bounded situations. For example, proofs requiring verbatim transcription or minor alterations from a template are adoptive in nature. This behaviour, they claim, would be typical of the *competent practitioner*. *Adaptive* learning requires higher cognitive processes such as creativity, reflection, criticism, and so on. Such behaviour would be typical of the *expert*.

They introduced these terms to draw a distinction between deep adoptive and deep adaptive learning. The former learner may be adept at complex tasks requiring a number of separate skills. The deep adaptive learner will be able to generate examples and proofs more independently showing insight, creativity and a higher degree of conceptualization. We talk loosely of *higher level skills* as those which typify an adaptive approach.

While a course is defined by its explicit curriculum it is the high-stakes assessments, the assessed curriculum, which really prescribe what is measured. Hence, the nature and purpose of the course are influenced strongly by the instruments used in summative assessments. If during such assessments we fail to ask the students to practice some technique, or to develop expertise—such as writing a mathematical proof—we can hardly complain afterwards that they have failed to do so. If we genuinely covet other skills, e.g. the use of computer technology, effective presentations, research and investigation, our schemes of work should reflect this. It is widely known that assessment drives what and how mathematics is learned and that students tend to adopt surface approaches to learning; Gibbs and Simpson (2004). Furthermore, mathematics education research indicates that students who adopt a deeper approach to learning are more successful; Dahlberg and Housman (1997). Rather than attempt to articulate what we mean by 'learning' in an abstract sense we instead concentrate on assessment items and what might be achieved when students become proficient in answering them.

3.5 Principles and tensions of assessment design

What constitutes 'good' mathematics assessment? This is the key concern, and to attempt to address this we articulate principles, and then acknowledge and discuss tensions which arise inevitably in practice.

Overarching principle: Assessment should reflect mathematical practice.

We shall try to be more specific in highlighting important aspects of mathematical practice. These aspects will inform parts of 'good' assessment of mathematical ability. Section 3.1 commented on the nature of mathematics, and fundamental to mathematics are problems.

Principle 1 Mathematicians try to solve problems.

Mathematical practice is concerned with finding solutions to problems, whether related to practical problems (applied) or internal to mathematics (pure). What is a 'solution'? In answering this we acknowledge the classical differences between computation and reasoning. In the former we seek systematic and efficient algorithms which, when applied to appropriate situations, are guaranteed to produce the correct answer. In the latter, it is the logical argument itself which is the focus of attention. In both cases it is the students' activity which is key. Indeed, Nunn (1911) claims that *'the point of immediate importance here is that mathematics is conceived not as a static body of "truths" but in the dynamic form of an activity'*. He cites three motivations as *wonder, utility* and *systematizing*: they all result ultimately in problems which need to be solved. If an assessment of students is to reflect practice, then problem-solving must be a key part of the instrument.

Ideally it should be clear when the correct solution has been found through straightforward internal checks, and we should expect to see evidence that students have performed such checks to confirm they believe their result. Assessments which ask a student to verify a given solution are an impoverished form of problem solving and are inherently less satisfactory.

Principle 2 Standard algorithms are both useful and important mathematical cultural artifacts in their own right.

Generally one thinks of algorithms as utilitarian, but they are also important cultural artifacts. Would anyone really argue that the Euclidean algorithm, the Sieve of Eratosthenes, or place value were not as significant as Leonardo da Vinci's *Mona Lisa* or Handel's *Messiah*? Of course they are also useful, and so once a problem has been formulated these standard techniques can often be applied. Where there is a choice of mathematical model for a particular problem, the choices made in how the problem is formulated have a direct bearing on the difficulty of the resultant equations. Hence, in genuine modelling situations awareness of what types of systems can be solved with the standard algorithms is key in deciding how to set up a model. Furthermore, experience with how to use these algorithms affects the precise ways in which the problems are modelled. A particular choice of coordinates can have a profound effect on the difficulty of the resultant algebra, and can make all the difference between success and failure in finding any solutions. Examples are given in Sangwin (2011b).

It is often important to assess (i) an understanding of when to apply such algorithms; (ii) an understanding of the details of how the algorithm works; and (iii) an ability to use the algorithm accurately and efficiently. Many standard algorithms can be automated, e.g. arithmetic on a calculator or more advanced operations on a computer algebra system. A consequence of this is that to assess (ii) and (iii), some summative assessments should be technology-free.

Details of the special cases are important; they reinforce the need to assess an understanding of when an algorithm is really applicable. For example, we should confirm whether students understand why 'division by zero' is forbidden, as the following question illustrates.

▼ **Example question 1**

Criticize the following argument. Suppose $a = b$ then $ab = a^2$, and so $ab - b^2 = a^2 - b^2$. Factoring gives $b(a-b) = (a+b)(a-b)$. Cancelling gives $b = a + b$. Since $a = b$ we have $b = 2b$. Hence $1 = 2$.

It is the justification which is as important as the answer in mathematics. Indeed, in pure mathematics the answer may be given or 'obvious', but the internal connections which give rise to the result might be fiendishly difficult to understand and hence justify.

Principle 3 Mathematicians justify their solutions. The outcome of mathematics is a correct chain of reasoning, from agreed hypotheses to a conclusion.

Note the difference here from empirical evidence. An explicit consequence of this principle is the need for students to correctly link multiple steps of calculation and reasoning. Hence,

Principle 4 Accuracy is important.

That is to say, in many situations an important goal is obtaining the correct answer to a particular question. The word 'accuracy' is also used here to include following a particular technique specified in a question, e.g. actually performing integration by parts rather than 'guess and check'.

Mathematics constitutes an intellectual sub-culture. Indeed, mathematics has its own history, folklore, and humour, Renteln and Dundes (2005). Like all human groups, mathematics has its own language including technical phrases, notation and ways of working which need to be appreciated. Some conventions are historical but persist, some are arbitrary, and some are very necessary.

Principle 5 It is important to acknowledge the place of conventions which should be distinguished from arbitrary definitions or logical consequences.

The last principle will be justified in more detail in Section 5.2 below.

Hence, we seek *problems* which allow *extended deductive reasoning* where *accurate* work can be demonstrated. The solutions should make use of *routine mathematical techniques* where students can adopt (or consciously ignore) *mathematicians' conventions*.

Clearly advanced students have ample opportunity to work on such problems. Is this a realistic goal for all levels, and from early years? We argue that traditional word problems possess many of the features of mathematical practice, and that they can be used at many levels in schools and at university. Hence, we argue that they play an important role in 'good assessment'. This view is widely supported.

> I hope I shall shock a few people in asserting that the most important single task of mathematical instruction in the secondary school is to teach the setting up of equations to solve word problems. [. . .] And so the future engineer, when he learns in the secondary school to set up equations to solve 'word problems' has a first taste of, and has an opportunity to acquire the attitude essential to, his principal professional use of mathematics. (Polya, 1962, Vol. I, p. 59)

These problems have been extensively studied by many authors, e.g. Gerofsky (1999) and Mayer (1981).

Since it is very difficult to discuss assessment of mathematics in broad and abstract terms we shall use traditional word problems to acknowledge and examine serious practical problems and tensions which both teachers and assessment designers need to address. In due course we shall also consider how we might automatically assess answers to questions such as these.

The following is a contrived but rather famous example of a traditional word problem.

▼ **Example question 2**

A dog starts in pursuit of a hare at a distance of 30 of his own leaps from her. The dog takes 5 leaps while the hare takes 6 but covers as much ground in 2 as the hare does in 3. In how many leaps of each will the hare be caught?

Not only are techniques and algorithms 'cultural artifacts', but these problems have also challenged and intrigued generations. A variant of this problem appears in *Propositiones ad acuendos juvenes* ('*Problems to sharpen the young*'), by Alcuin of York, c.800 AD, see Hadley and Singmaster (1992). It has been used continuously ever since, e.g. in Tuckey (1904, Ex 65, (44)). It may even be older:

> This is the first known European occurrence of this 'hound and hare' problem. There are many examples of overtaking problems, often more complicated, in the *Chiu chang suan ching* of c.150 BC. Overtaking problems also occur in the Indian and Arabic literature about the time of Alcuin. (Hadley and Singmaster, 1992, p. 115)

However, some traditional word problems assume a certain level of cultural knowledge. It cannot be taken for granted that contemporary students, particularly those from urban areas, know with confidence what a 'hare' is. Whatever the purpose of word problems, in a mathematics class they are certainly not intended as a test of such cultural knowledge. This, and other difficulties with word problems, are discussed in, for example, Cooper and Dunne (2000) and Mason (2001). We reasonably expect all students in post-compulsory education to be familiar with the SI system of metrology, time and with currency. However, students may not be familiar with the rules of sports, or other games. Hence, when used as summative assessments due consideration needs to be given to ensure that all students are treated equitably.

The following question is ubiquitous:

▼ **Example question 3**

A rectangle has length 8cm greater than its width. If it has an area of 33cm^2, find the dimensions of the rectangle.

Interpreting such problems to derive the correct equations is far from easy: problems involving ratio, for example as rates, appear to be particularly difficult.

▼ **Example question 4**

Alice and Bob take 2 hours to dig a hole together. Bob and Chris take 3 hours to dig the hole. Chris and Alice take 4 hours to dig the same hole. How long would all six of them take by working together?

The temptation is to model the work of Alice and Bob as $A + B = 2$, rather than $A + B = \frac{1}{2}$. There are real difficulties in reaching a correct interpretation, and hence in moving from a word problem to a mathematical system which represents it. Consider a problem related to, but different from, 4 in which pairs of people 'walk into town' rather than 'dig a hole'. Similar conceptual difficulties occur with concentration and dilution problems.

We argue that moving from such word problems to mathematical systems constitutes the beginning of mathematical modelling and hence is a valid component of practice. Principle

1 asserts that problem-solving is key to mathematics. If mathematicians solve problems, then problem-solving should be assessed. However, genuine problem-solving is a difficult and time-consuming activity; see Badger et al. (2012). Furthermore, problem-solving can be easily replaced with memory. This provides us with our first tension.

Tension 1 *Seen vs unseen*

Given that problem-solving is difficult, it is perfectly reasonable, and significantly challenging, to ask students to abstract information from a word problem; formulate it using conventions; recognize this as a standard case for which a known technique is applicable; and accurately solve the equations. Indeed, the whole point of systems such as place value, algebra, and calculus is to group similar problems in a systematic way to prevent everything becoming a puzzle. It is intriguing and instructive to look at the *ad hoc* methods used to solve quadrature problems and compare these with the modern equivalents which benefit from algebra and calculus; e.g. Edwards (1979). However, by the very nature of such sophisticated intellectual tools as algebra, for each individual recognizing that a word problem can be modelled by a particular system of equations appears to require a high level of fluency. Being able to select and correctly use standard techniques comes only with practice. This is important, since working memory is limited, and sufficient familiarity and fluency are needed to ensure an automatic response. This frees the mind to think at a more strategic level. This leads us to ask two questions:

1 What is the role of memory in mathematical practice?
2 What is the interplay between memory and recognition?

Authors such as Gattengo (1988) have addressed these questions by thinking about *awareness* and *attention*.

> Awareness alone is like knowledge without skill: nothing actually happens. Learning involves both educating awareness and using that awareness to direct behaviour, which can be trained by developing habits. (Mason and Johnston-Wilder, 2004, p. 63)

At the most basic level there is some agreement on the existence of a subtle interplay between *remembering* and *understanding*.

> There are certainly some things in mathematics which need to be learned by heart, but we do not believe that is should ever be necessary in the teaching of mathematics to commit things to memory without at the same time seeking to develop a proper understanding of the mathematics. (Cockcroft, 1982, §238)

Furthermore, during assessment design, tension 1 manifests itself in deciding what constitutes a 'trick'. What, at first sight, constitutes a trick is often just a useful mental move, worth remembering for potential future use in similar problems. When to use it is both a function of memory and recognition. For example, is recognizing $\sin^2(t) - 1$ as *the difference of two squares* a trick? Is the following a 'trick question'?

▼ **Example question 5**

Calculate
$$\int \frac{1}{\cos^2(t) + \sin^2(t)} \, dt.$$

The 'trickiness' of a particular move is a function of the familiarity of the individual with the material, and their ability to make connections. Whether or not this is a trick question, it is sometimes impossible to avoid rarely used, or unique, mathematical moves. Writing $\ln(x)$ as $1 \times \ln(x)$ before integrating by parts is a classic trick. Here, memory probably has a larger part to play than genuine creativity.

Another consequence of tension 1 is that it is almost impossible to evaluate assessments in isolation from the stated and enacted curriculum. Each question, such as those given here as examples, is posed in a complex context which is difficult to describe accurately.

A related tension is what Brousseau (1997) calls the *didactic tension*. Essentially, if I tell you the answer then I rob you of the satisfaction of discovering it for yourself. So, to what extent should students be allowed to struggle, and how much should they be told explicitly? This leads us to another tension.

Tension 2 *Guided vs independent work*

If word problems are 'abstract' and decontextualized then they appear contrived, divorced from reality, and even ridiculous. However, the analysis of Cooper and Dunne (2000) found that social class was a significant factor in determining children's performance. In particular they report that *'working class students performed equally as well as their middle class counterparts on decontextualized test items but struggled on "realistic items" which were embedded in everyday contexts'*.

Furthermore, pure mathematics seeks a complete understanding, and only a subset of the possible equations or parameter ranges may occur in practical problems. While word problems may have a place in a modelling course, they can also distract students with irrelevant detail or 'noise'. Hence, we have a further tension.

Tension 3 *Abstract vs context*

Seeing the practice of mathematics itself as solving problems through modelling, and thus understanding the satisfaction of dispatching the routine steps accurately and efficiently, may act as motivation for students to undertake the repetitive work needed for the acquisition of skills. Alternatively, students may prefer to learn topics initially in an abstracted form and practise these to gain fluency before trying to solve problems which require their use.

> Drill in mere manipulation is necessary at every stage in school algebra. That this should be thorough, so far as it goes, will be admitted by all teachers, but it should in the main

be given *after* its necessity in applications has been perceived by the pupil and not *before*; also, it should not be carried further than is needed to ensure facility in these applications. (Tuckey, 1934, p. 10)

If we accept principle 3 combined with 4 then we argue that it is the product of mathematics which is most highly valued. But what about the process? That is to say, what is the method and reasoning? There is certainly a tension here.

Tension 4 *Product vs process*

In assessment design this tension manifests itself in assigning appropriate weight to method marks, and the extent of 'follow-through marking', in which the effects of a mistake are ignored in subsequent working. Follow-through marking of extended unstructured problems becomes very problematic, particularly when there are multiple mistakes, and this might make assessments impractical.

Are we going to reward students for *selecting* a method, or are we going to be explicit in telling them which method to use? This of course depends on which outcomes are being assessed. What is the role of guessing? For example, mathematics questions usually have tidy answers, so in Example question 3 we might start by factoring $33 = 11 \times 3$, and then we notice that the difference is 8, just as required. There is no need to write down a quadratic, nor to solve it. In some areas guessing is almost elevated to a technique; for example, symbolic integration of elementary functions. In others, the form of a solution is guessed, and on this basis coefficients or parameters inferred; for example, in assuming that a differential equation has a solution of the form $e^{\lambda t}$.

We have chosen to concentrate on word problems in this section to illustrate the tensions which arise when designing assessments which try to conform with our principles. In the following example, however, process is everything.

▼ **Example question 6**

Differentiate $f(x) = x^6$ with respect to x
(i) as a product $x^2 \times x^4$ then as a product $x^3 \times x^3$;
(ii) as a composition of x^2 with x^3, i.e. $(x^2)^3$;
(iii) as a fraction $\frac{x^{6+n}}{x^n}$.

The above question deliberately extends a single step differentiation to include the product rule, chain rule and quotient rules. It may be possible to devise many similar assessments in which a student is questioned at length about their solution to an extended problem on which they have been working for some time, e.g. as a *viva voce*. While they may have had access to reference materials originally, now we can ask them about their understanding and application of these to their new situation. Limitations on resources probably make such examinations practical only for the few doctoral candidates. Hence, assessment designers need to reach a compromise to ensure that it is practical to achieve sufficient reliability within the available resources. There is certainly a tension between validity and practicality, and computer aided assessment may well have a place here.

Tension 5 *Validity vs practicality*

Extended project work has proved to be very difficult to assess: impersonation or plagiarism are serious practical problems which invalidate assessment and cannot be ignored. For this reason the unseen and timed examination undertaken in closely controlled environments seems likely to remain the primary instrument for high-stakes summative assessments for the foreseeable future. However, even here strict precautions are necessary when curricula, such as those of the International Baccalaureate, are studied and examined in multiple time zones, and where the potential exists for candidates to see and benefit from discussion of questions which are posted online.

In Section 3.3 we defined the purposes of assessment. For pragmatic reasons, an individual instrument may be used for a number of different purposes. For example, an 'exercise sheet' may have a primarily formative function, and students' work could be marked against model solutions with comments provided drawing students' attention to correct and incorrect working. The student's achievement is then reduced to a single numerical mark which is itself a crude formative measure. This mark could also be recorded for summative purposes as a contribution to the overall module/course mark. Qualitative comments could be aggregated as an evaluative assessment to inform subsequent teaching (*quality enhancement*). Marks for each separate item could form an evaluative assessment to ensure a team of teaching assistants are marking in a consistent and fair manner: *quality audit*. It is argued by, for example, Wiliam and Black (1996) that it is the use to which the outcomes of an assessment are put which has a greater bearing on the purpose of the assessment than the form of these outcomes. For example, numerical outcomes can be formative, summative, or evaluative. It is also interesting to note that the verbs 'mark' and 'grade' are commonly used by teachers to refer to the process of traditional assessment. Furthermore, strong messages are communicated to students by the choices made for assessment: a '*reward for sustained achievement*' needs to be balanced against an '*opportunity to learn from mistakes*'. Alternatively, there are periodic complaints of '*teaching to the test*'. If the test is right, then what is wrong with teaching to the test?

> [...] the *implemented* (or enacted) *curriculum* will inevitably be close to the *tested curriculum*. If you wish to implement the *intended curriculum*, the tests must cover its goals in a balanced way. (Burkhardt and Swan, 2012)

This provides us with another tension:

Tension 6 *Summative vs formative*

Many teachers will also recognize the difference in levels of engagement in general teaching situations between purely formative work and course work which contains a small summative component to it. This problem seems inescapable.

If our broader goal is education rather than technical training then we argue that for the purposes of summative assessment it is not sufficient only to ask students to apply standard techniques to predictable equations. They are not 'problems' in our sense, but practice exercises. Exercises are perfectly appropriate in the formative stages of learning, but they do not constitute the product of practice. Hence, the form of a formative exercise might differ significantly from a summative assessment. There is a spectrum of problems from entirely routine situations to those where a particular insight is applicable only in one case. In line with the broader goal of the course, a range can be selected.

Principle 3 asserts the importance of extended chains of reasoning. If this is accepted then these should be assessed. Simple, routine, exercises are described by Gardiner (2006) as *single-piece jigsaws*.

Tension 7 *Structure vs freedom*

In assessment design this tension manifests itself in how much structure to provide for students. This structure might be explicit instructions to solve a problem using prescribed steps, or it might be implicit in the allocation of marks for each section. Students become adept at drawing inferences from the allocation of marks, which may reduce the validity of the assessment. In this book we shall see many examples of CAA systems in which problems are broken down into predefined steps. There are fewer systems which accept a free-text response from students for automatic assessment.

Word problems immediately turn a single-step mathematical exercise into a multi-step chain of reasoning. In Example 2, let l be the number of leaps taken by the dog. The problem reduces to solving

$$l = \frac{6}{5} \times \frac{2}{3}l + 30.$$

This is a simple linear equation, but it can be arrived at only by careful work on the part of the student. In Example 3 the student is free to choose either the length or width of the rectangle as a variable. Ignoring the particular letter used for the variable, this choice results in one of two different equations, i.e. $x(x + 8) = 33$ or $x(x - 8) = 33$. One solution must be discarded as 'unrealistic': a valuable critical judgement by the student. However, as we have discussed, the solution to this problem might be found by inspection.

3.6 Learning cycles and feedback

Assessment with feedback occurs in *cycles*. The work of Kolb (1984), for example, proposed a cycle describing the actions of someone faced with a learning task. They undertake a cyclic process of (i) abstract conceptualisation; (ii) active experimentation; (iii) concrete experience; and (iv) reflective observation. Other authors, e.g. (Laurillard, 2002, Figure 6.1, p. 113), have further elaborated these basic ideas into complex iterated cycles of events between the student's and teacher's actions. In all such models of learning, feedback plays a central role.

It is axiomatic in education that 'feedback promotes learning'. However, the research evidence on this matter appears less clear-cut. The meta-analysis of Kluger and DeNisi (1996) examined about 3,000 educational studies. Of those which met basic soundness criteria they found that over one third of feedback interventions *decreased performance*—a counterintuitive and largely ignored outcome. They concluded, not surprisingly perhaps, that it is not feedback itself but the nature of the feedback which determines whether it is effective. In particular, feedback which concentrates on specific *task-related* features and on how to improve is found to be effective, whereas feedback which focuses on the *self* is detrimental. A low end-of-test summary mark—hardly a specific form of feedback—may be interpreted as a personal and general comment on the ability of the student, whereas detailed feedback on each task points to where improvement can be made. Shute (2007) concluded as follows.

> Formative feedback might be likened to 'a good murder' in that effective and useful feedback depends on three things: (a) *motive* (the student needs it), (b) *opportunity* (the student receives it in time to use it), (c) *means* (the student is able and willing to use it). However, even with motive, opportunity, and means, there is still large variability of feedback effects on performance and learning, including negative findings that have historically been ignored in the literature (see Kluger and DeNisi (1996)). (Shute, 2007, p. 175)

During automatic CAA feedback is potentially instantaneous, whereas traditional assessment of written work often involves a delay. This may be significant, of the order of days or perhaps even weeks. By this time the student may have forgotten the details and have more pressing new work to complete. For high-achieving students Shute (2007) recommends using delayed feedback, facilitative feedback and that verification feedback may be sufficient. However, for low-achieving learners Shute (2007) recommends the use of immediate feedback which is directive (or corrective).

In mathematics, providing students with a 'worked example' or 'model answer' is very common. They may be given as preparatory examples or in response to a student's answer to a particular problem once assessment has taken place. They certainly provide task-specific information, and students find then very useful. Borrowing from engineering they might be termed *open loop control*, because they are not based on, and do not respond to, the student's activity directly. Therefore, they cannot be considered true feedback in the strict sense of the word.

> Feedback is a method of controlling a system by reinserting into it the results of its past performance. If these results are merely used as numerical data for the criticism of the system and its regulation, we have the simple feedback of the control engineers. If, however, the information which proceeds backwards from the performance is able to change the general method and patterns of performance, we have a process which may well be called learning. (Wiener, 1968, p. 56)

The origins of formative feedback lie in the 1930s, from constructivist and cognitive theories of learning through cybernetics (Roos and Hamilton, 2005) and system control (Sadler, 1998). The kinds of feedback generated by CAA which we describe here can respond to the student's actual answer, and therefore has the potential to help them improve on the task. If individual tasks are seen by students to represent more general situations, then the feedback may well promote genuine learning.

When using CAA, the teacher begins by devising a task *and* decides at the outset what feedback should be provided. Without trials, it may be quite difficult to predict what mistakes are likely and thus to provide helpful or even sensible feedback. This is clearly a challenge in assessment design that is not present in the traditional learning cycle. It is for this reason that tools for subsequent detailed analysis and diagnosis are so useful to the teacher for improving and validating feedback. While the immediacy of such feedback is a feature of all CAA, the ability to generate tailored feedback based on mathematical manipulations of the student's answer is one of the main strengths of CAS-enabled CAA support which we will examine in detail. We emphasize that CAA shifts the workload from the assessment to one of a preemptive nature.

A potential unintended consequence of CAA, as described in this book, is a possible *reduction* in the ability of some students to identify their own mistakes. They may become used to having it done for them immediately, much in the same way that many students now struggle with arithmetic as they are used to their calculator doing it for them. While we have no evidence for this, any changes in the form of assessment need to be carefully considered. Indeed, our experience with CAA suggests that students actually welcome the ability to practise with randomly generated questions and do read the worked solutions in detail.

As with many of the topics in this chapter, feedback is a complex business. CAA certainly provides rapid, criteria-based feedback. Detailed feedback is time-consuming in providing for each student by hand, and the teacher faced with a large cohort may simply not have sufficient time to provide the quality of individual feedback they would, under ideal circumstances, wish to provide. Automatic feedback in CAA can be scaled to very large groups without costs in terms of staff time growing linearly with group size. A counterbalance to this is the loss of a medium-term review of work which would normally take place when marked work is returned to a student. Now, there is no opportunity inherent in the learning cycle for the student to revisit and review their work at a subsequent time, perhaps a week later.

Another point to note is that in this new cycle the teacher is now somewhat detached from the process. Teachers do not have the same imperative to review the progress of the individual student, or the entire cohort, as they would if marking work by hand. However, one strength of the automated system is being able to correlate and process the entire cohorts' responses. Such a facility offers the teacher the ability to spot common mistakes routinely and quantify these. In the traditional approach, all that was possible was to gain a 'sense', albeit an often accurate one, of what mistakes were common without a formal action research project.

3.7 Conclusion

This chapter has attempted a very difficult task; namely, to discuss what it means to learn mathematics and how we might assess such learning. It has also defined terms used to describe aspects of the assessment process. There are two important strands to mathematical activity.

1 *The use of routine techniques*
   This includes recognition and the reduction of problems to cases for which a standard algorithm is applicable.
2 *Problem-solving*
   For the solver this involves elements of novelty which demand creativity and often personal struggle. Sometimes solutions are useful for solving similar problems, while in other cases the argument is somehow unique.

The two strands are inseparable. Problem-solving may be replaced by memory or research (with a small 'r', i.e. looking up the answer). Without sufficient practice, recognition is impossible and all mathematical questions become problem-solving, which is inefficient and causes problems in recording and communicating mathematics. What is the point of simply being good at technique if we do not apply these techniques to solve problems? Each teacher has to address the tensions articulated in this chapter for themselves in the context of their curriculum. For each student, gaining a mathematical education is the personal process of reinventing the wheel, and thereby joining an international community of practice.

4

Mathematical question spaces

Student: With these online quizzes, whenever I finish a quiz, I get the option for a different new version.

Myself: Well, the questions are randomly generated.

Student: How many versions are there? How do I know when I have *finished?*

Myself: As I said, since the questions are random, there are infinitely many different combinations.

Student: So that means I cannot ever finish!

Myself: The numbers are random, but the questions are from a template. You have to decide what each question is really about. You are 'finished' when you are confident you could answer any version correctly when asked. The numbers *do not really matter* do they? What is this question actually asking?

This conversation is not entirely fictitious. We had been running CAA for a number of years, and this included practice of routine methods. Students could repeatedly obtain randomly generated quizzes. To us it was obvious what the questions 'were about', and why we had chosen to randomize them in particular ways. But to this student who came to my office it was not at all obvious. To him an important component of *success* was completing all the work set by the teacher. Actually, since we had appreciated early on the importance of being conservative about how we randomized the questions there were only a few different versions.

This chapter examines the process of randomly generating versions of mathematical questions for CAA. In doing this we examine not only a single mathematical question, but how such questions are linked together into coherent structured schemes. Using schemes of questions is a widely used technique when studying mathematics.

It is interesting to note that the need for teachers to generate many equivalent mathematical questions arose not with CAA, but with teaching itself. Plympton 322 is a clay tablet measuring 12.7×8.8cm. It was written in the ancient Iraqi city of Larsa (modern Tell Senkereh, $31°14'$N, $45°51'$E) in the mid-eighteenth century BC. Of this, Robson (2001) wrote: *'Plimpton 322 has undoubtedly had the most extensive publication history of any cuneiform tablet, mathematical or otherwise.'* Her analysis continues:

> On balance, then, Plimpton 322 was probably (but not certainly!) a good copy of a teachers' list, with two or three columns, now missing, containing starting parameters for a set of problems, one or two columns with intermediate results (Column I and perhaps a missing column to its left), and two columns with final results (II–III). (Robson, 2001)

Hence, the teachers of the scribe class of ancient Babylon faced problems similar to those faced by authors of CAA today.

4.1 Why randomly generate questions?

Why do we want to randomly generate questions? In computer aided assessment we do not have any handwriting evidence. If every student got an identical quiz, as in the traditional paper-based approach, there would be no way to detect plagiarism. Of course, we still have the different problem of impersonation, but students are unlikely to be willing to undertake this on a large scale if the questions are different. We do know that impersonation happens: having the name of the student visible on every screen enables an invigilator in a high-stakes assessment to confirm identity. The twin difficulties of impersonation and plagiarism are present in all assessment regimes, including traditional approaches. It ranges from the petty intellectual dishonesty of asking a peer for the *crucial step*, and then passing off the results as one's own, right up to wholesale copying of entire projects. The extent to which a teacher needs to prevent students behaving in these ways depends on whether the assessment is summative and the stakes of these assessments. Randomization has for some time been used to reduce these problems, but we do not wish to dwell on these difficult issues here. Teachers need to consider the level of trust they have with their students, and the stakes in their circumstances. Actually, randomly generating questions and giving different versions to each student creates opportunities not present with a fixed paper worksheet, and we shall concentrate instead on these.

The first important reason is that distinct but equivalent questions may be used for practice. We shall try to examine what *equivalent* means. Furthermore, giving each student a different version would allow them to collaborate without fear of plagiarism. They can discuss the general method, confirm that it works in each individual case, and provide evidence of their work. This collaborative effort could alert them to precisely the structure underlying the question-generation process and so help them understand the range of applicability of the particular technique needed in this case. As one of our students commented in their feedback evaluations to the AiM system:

> *The questions are of the same style and want the same things but they are subtly different, which means you can talk to a friend about a certain question but they cannot do it for you. You have to work it all out for yourself, which is good.*

Notice here the student voices the opinion that the questions '*want the same things but they are subtly different*'. We shall address exactly this issue by examining equivalent mathematical problems in some detail.

4.2 Randomly generating an individual question

As a simple example consider the following exercise:

$$\text{Solve } ax^2 + bx + c = 0. \tag{4.1}$$

A first approach would be to randomly generate integer coefficients $a \neq 0$, b, and c, and to substitute these into an expression template $\Box x^2 + \Box x + \Box = 0$. We might consider creating a 'question space' by indexing the individual instances with coordinates $(a, b, c) \in \mathbb{R}^3$. Clearly, there are some 'subspaces', such as the subspaces of mathematically possible questions. In our example, the subspace satisfying $b^2 \geq 4ac$ characterizes the question subspace with real solutions.

This approach is not sensible, for both technical and educational reasons. At a purely technical level, it is likely that we will display unsatisfactory equations such as $1x^2 + -8x + 15 = 0$, which do not conform to notational conventions. It is possible to work around these problems by having, in this case, four templates. That is to say one for each of $\Box x^2 + \Box x + \Box = 0$, $\Box x^2 - \Box x + \Box = 0$, $\Box x^2 + \Box x - \Box = 0$, and $\Box x^2 - \Box x - \Box = 0$ and ensuring validity by only generating non-negative integers for b and c and an integer $a \neq 0$. This is artificial and unsophisticated, and quickly leads to extra workload in maintaining multiple copies of almost identical questions for each case. It is surprising how many CAA designers avoid even the most rudimentary CAS to support this task.

The most expedient solution is to begin with integer roots, α and β say, and to have computer algebra expand out the expression $a(x - \alpha)(x - \beta)$. There are many advantages to this approach. We ensure that the resultant polynomial really does factor over the integers, whereas just randomizing the coefficients of (4.1) does not. By judicious choices of the roots we can control the cases a student sees, e.g. distinct/repeated roots, the signs of the roots, the magnitude of the integers (which may test the extent of fluent mental arithmetic), and so on. Even in cases where we want complex roots it is more sensible to begin with the answer and work backwards in this way. By having the roots as explicit variables in the CAA item at the outset we can manipulate them with computer algebra to produce the steps in any worked solution for the student, or to generate any potential answers which occur as a result of common mistakes/misconceptions. This particular example can (and indeed has) been automated without CAS. In this case, CAS is the natural tool and providing the teacher with access to CAS tools within CAA is invaluable. It clearly enables them to go much further than this example. We shall return to this issue and discuss the characteristics of such a CAS in more detail in Chapter 6.

We have described random numbers when in fact we mean *pseudo-random numbers*. It is quite important in practice when a student makes a subsequent visit to a CAA system that they get the *same version* of the question that they had originally. This can be achieved by storing a version of the question in a database, or by storing a *seed* from which a pseudo-random sequence is generated. Perhaps the simplest method of creating such a sequence is using a linear recurrence relation. For example, choose $\alpha := 2^{34} + 2^{17} + 4 + 1$ and prime $p = 2^{61} - 1$ and then generate the sequence (r_n) by:

$$\left.\begin{array}{rcll} r_1 & := & \text{seed} & \mod p, \\ r_{n+1} & := & \alpha r_n & \mod p. \end{array}\right\}$$

For more information on how the choice of α and p effect the statistical properties of the sequence, see Knuth (1969).

Another method of achieving randomization within a quiz is to make a selection of one from a list of stored items for each quiz item. Some systems, e.g. AiM, allow questions to be labelled with keywords so that a selection is made from those questions with a certain keyword. This technique is implemented in many CAA systems, and it allows much more flexible randomization to be implemented. Of course, this has nothing to do with structure within mathematical questions. Furthermore, selecting from a list of possible items does not preclude internal randomization as described above.

Regardless of how we generate this randomization, a crucial distinction, when considering a mathematical question, is whether or not one cares about the answer. With many questions, no one cares about the actual answer. The purpose of the question is either to (i) practise some technique, or (ii) help build or reinforce some concept by prompting reflective activity. In other cases the purpose of the question is to obtain the answer. The question itself is a prototype of a practical problem which may be encountered, and hence this result may be useful. For now, we shall assume that the teacher wishes to assess students' attempts at problems for which there exists a well-known and entirely standard technique. It is precisely these kinds of question where CAA appears to have immediate application.

It is interesting to note that some textbooks use the word 'example' in place of 'exercise'. Indeed, just as we might consider $\sum_{n=1}^{\infty} \frac{1}{n}$ to be an interesting example of a mathematical series, we consider a full worked solution to be a mathematical 'example'. It exemplifies the successful use of a logical argument, particular technique, procedure, or algorithm, rather than exemplifying an object (i.e., something which satisfies a definition) or as a counterexample. Returning to (4.1), let us consider one particular version.

▼ **Example question 7**

Solve $x^2 - 5x + 6 = 0$. (4.2)

We can factorize the left-hand side of (4.2) to give

$$(x-2)(x-3) = 0.$$

Recall that if $AB = 0$ then $A = 0$ or $B = 0$. Hence, $x - 2 = 0$ or $x - 3 = 0$. Rearranging these equations gives $x = 2$ or $x = 3$.

We might argue about the level of detail provided in this worked solution. For example, we have chosen here to factor the polynomial in a single step. Nevertheless, it is clear, once the roots have been chosen, how this *example* might become a template for random versions.

To help us consider what such an exercise does 'exemplify' it is entirely appropriate that we consider such exercises in the light of the work of Watson and Mason (2002a) who developed the concept of an *example space*. An example space is taken to be the cognitive domain possed by the student, rather than some intrinsic mathematical space. We seek to develop a dual notion: that of mathematical *question space*. Just as with example spaces, the notions of the *dimensions of possible variation* and *ranges of permissable change* in any question space appear to be very useful. Each dimension of possible variation corresponds to an aspect of the question which can be varied to generate a collection of different question instances. The range of permissible change is more problematic. 'Permissible' may be taken to indicate the strict mathematical criteria of well-posedness, or may be used in a pedagogic sense. Given our educational context, a *question space* is taken to be the collection of instances which are educationally equivalent. While the student is likely to be aware only of the task in hand—the question instance—to the teacher this instance actually represents the question space and hence the underlying generality.

Let us consider some possible dimensions of variation in (4.1). Clearly the *roots of the quadratic* are one dimension of variation, and we probably consider only small integers. What happens if we have a repeated root? Is the difference of two squares significantly different for our students or will our students be thrown by the lack of an x term in (4.1)? We have used x as the variable. Might we select a letter from a list to provide variety? In (4.1), as in many standard textbooks, the equation is already provided in *standard form* or *canonical form*. What about an equation such as 'solve $x^2 = 5x - 6$'? In this latter example, we might decide to have a new first line in our worked solution such as '*Writing the equation in a standard form $ax^2 + bx + c = 0$ gives*. . .'. Does this *change the question*?

In practice, a simple and effective way to establish the limits of randomization is to consider carefully the *fully worked solution*. This will consist of a number of steps, and quite what constitutes a 'step' in working will be highly dependent on the group of students for which the question is intended. By constraining the randomly generated parameters in such a way that the worked solution is invariant we ensure the students can solve the problem using the model solution with which they are provided. A key concern is whether this work actually promotes the reflective activity which alerts the student to this potential generality. This depends on how the question itself is actually used, and the role of the teacher is key here.

Let us remind ourselves, at this point, of Babbage's difficulty in 'slicing a pineapple', which is quoted at the start of Chapter 3. The notions associated with 'question space' can be applied in those situations where a teacher has a worked solution in mind. This does not mean that rigorous alternative solutions do not exist, or that this notion can be applied to every type of mathematical question. Proofs, which often contain an element of generality anyway, are not so easily examined. However, when trying to randomly generate mathematical questions for routine practice, then these notions are very helpful indeed to the question author in considering the extent to which effective randomization may take place. Some examples are given in Section 9.1.

The practice of some technique could be seen to be the repeated completion of question instances from a particular question space. For example, it would be possible to provide practice by creating a quiz which contains half a dozen different versions of question (4.1). Such repetition on its own is unlikely to constitute particularly effective practice. The work of Ericsson *et al.* (1993) stresses that to develop expertise *deliberate practice* is required and that mere repetition is not sufficient. For optimal learning and improvement, practice tasks must take into account the learner's current level of understanding and performance. The role of the teacher is essential in sequencing these tasks appropriately and monitoring performance to decide when to introduce more complex and challenging tasks. However, *adaptive testing* seeks to encode such decisions into computer aided assessment. Equally students must also recognize that deliberate practice is *effortful* and can be sustained only for a limited period of time without leading to exhaustion. This point of view is supported by Hewitt (1996).

> Furthermore, repetition is designed to help students stand still. [. . .] It is not repetition in order to stay still but practice by moving forward and progressing in their learning. This is what I call *practice through progress*—practising something while progressing with something else. [. . .] And repetition takes up considerable time, leads to boredom and lack of progress, and is mainly concerned with short-term 'success'. (Hewitt, 1996)

A selection of questions usually shows progression through a sequence of slightly different cases. Each of these will be consciously different, and so will be instances from different question spaces. Recall that the notion of 'educational equivalence' defined here is rooted in the idea of invariance of the worked solution. Different cases have different worked solutions. To generate such a structured sequence will require the variation to be rather more constrained than would appear necessary at first sight. We shall consider structured sequences of questions in the next section.

In many situations the mathematical aspects of a question space may be far from obvious. For example, when does Question 4.1 have rational solutions? For polynomials of higher degree such questions constitute an important mathematical topic in their own right. Actually reverse-engineering questions is itself a rather interesting and undervalued mathematical activity. In particular, some very interesting mathematical problems arise when trying to generate 'nice' questions. Even finding 'nice' cubics, i.e. a cubic polynomial with rational roots and rational coordinates for the stationary points, leads to interesting mathematics; see Johnson (2011). For some examples in linear algebra see Steele (2003), for differential equations see Cerval-Peña (2008), and for graph theory see Ruokokoski (2009). Even discerning which instances of a question are possible might be a challenging problem. Indeed, characterizing all those routine problems with particular mathematical characteristics is a difficult task, especially at university level. This, of course, is a challenge which could be passed on to students.

4.3 Linking mathematical questions

We begin our examination of mathematical questions with a sequence of simple questions from Tuckey (1904). This small, unassuming volume consists of 178 pages. There is no text or worked examples; instead, simply sequences of problems. '*These examples are intended to provide a complete course of elementary algebra for classes in which the bookwork is supplied by the teacher*'. One such sequence is shown in Figure 4.1.

> Draw the graphs of:
> (1) $y = x^2$.   (2) $y = -x^2$.   (3) $y = 2x^2$.
> (4) $y = x^2 + 2\cdot 5$. (5) $y = (x-1)^2$. (6) $y = (x+2)^2 + 1$.
> (7) $y = x^2 + 4x + 6$.   (8) $y = x^2 - 3x + 1$.
> (9) Write out a general statement of the difference between the graphs of $y = x^2$ and of $y = \pm a\{(x-b)^2 + c\}$.

Figure 4.1: A sequence of simple questions. (Tuckey, 1904, p. 62)

This sequence of questions is highly structured, and this example has been included here because clues to this structure are revealed in the unusual final synoptic question. Question 1 provides a base from which comparisons can be made. Questions 2 and 3 transform the y-axis. Question 4 is a vertical shift, question 5 is a horizontal one, and question 6 involves both. Questions 7 and 8 also require shifts, although some simple algebra is required to reveal precisely what these are.

It appears clear that the purpose of such a sequence of questions is to develop concepts rather than obtain an answer or simply to practice technique for its own sake. However, the concepts are actually developed *through practice*. That is, what might appear naively as practice turns out to be rather interesting. In asking the students to complete such a sequence of exercises the teacher is asking for the trust of the students. This trust forms part of the *didactic contract* described by Brousseau (1997). The responsibility of the teacher, in this contract, is to provide materials which end up being stimulating, and with which students can engage so as to learn. Notice here the similarity with the notion of *subordination* described by Hewitt (1996). He suggests that learning takes place effectively as *'practice through progress'*, and that a skill is *subordinate* to a task in the following sense:

> I say that skill, A, is subordinate to a task, B, only if the situation has the following features.
> (a) I *require* A in order to do B. (This may be an existing necessity or can be created through the 'rules' of a task)
> (b) I can *see the consequences of my actions* of A on B, at the same time as making those actions.
> (c) I do not need to be knowledgeable about, or be able to do A, in order to understand the task, B. (Hewitt, 1996)

Note that the final question 9 shown in Figure 4.1 reveals the internal structure of the exercises. There is a tension here, since including such a synoptic question may make little sense for a student who has struggled with questions (1)–(8), and has little work of merit from which to form a coherent synopsis. Immediate feedback from CAA can potentially aid such students by ensuring that they eventually have a correct answer from which to form a synopsis.

As we have already seen, the domains of possible variation are certainly more complex than simply changing a coefficient in a term. For example, in question 7 of the problem set shown in Figure 4.1, the question is an instance of a quadratic with no real roots, for which the completed square form is tractable. An instance of such a question would probably be generated as an expansion of $(x - a)^2 + b$, where a is a small integer, and $b > 0$ is a small integer. Hence, a particular dimension of variation certainly does not correspond to the direct variation of a coefficient in a question instance. As a result, to implement randomly generated instances in CAA sophisticated tools are necessary, such as a CAS.

Clearly, here it is easy to identify how the dimensions of variation affect the question instances, but it is unlikely that such an algebraic clarity will be evident in many situations. Equally, there is nothing to suppose that a dimension of variation will be algebraic at all. Variation could include which variable is used, the dimensions and orientation of geometric shapes, or the adjectives used in a word problem. Furthermore, there are many situations when a parameter will remain within a question, perhaps to suggest to the student that there is a range of permissible and 'essentially the same' examples encapsulated within one question. It is possible in some circumstances that a question space will contain only one instance. For example, in Figure 4.1, question 1, there may be no reasonable alternatives, and the question space consists only of the instance 'Draw the graph of x^2'.

4.4 Building up conceptions

It is clear that constrained variation of similar problems could be used to generate questions in a structured way for practice. However, there are *sequences of questions* in which the use of variation can potentially alert students to structure in mathematics. For example, consider the following from Krause, 1975, Chapter 1. Note that the *taxicab distance* d_T between points $A = (a_1, a_2)$ and $B = (b_1, b_2)$ in the plane is defined as

$$d_T(A, B) = |a_1 - b_1| + |a_2 - b_2|.$$

▼ Example question 8

---

3 For this exercise $A = (-2, 1)$. Mark A on a sheet of graph paper. For each point P below calculate $d_T(P, A)$ and mark P on the graph paper.

   a) $P = (1, -1)$ b) $P = (-2, -4)$ c) $P = (-1, -3)$
   d) $P = (0, -2)$ e) $P = (\frac{1}{2}, -1\frac{1}{2})$ f) $P = (-1\frac{1}{2}, -3\frac{1}{2})$
   g) $P = (0, 0)$ h) $P = (-2, 2)$

4 a) Find some more points at taxi distance 3 from A.
   b) Graph the set of *all* points at taxi distance 3 from A; that is, graph $\{P | d_T(P, A) = 3\}$.

This sequence begins with a number of routine practice problems which require the student to plot points and evaluate the distance from $A = (-2, 1)$ using the new metric d_T. Random versions of such a sequence could easily be generated automatically, but what structure should we aim to maintain?

Although not as immediately evident as with the questions of Figure 4.1, Example question 8 part 3 above has an important structure which is examined further. The answers to parts (a)–(f) inclusive have two important mathematical properties: (i) co-linearity, and (ii) constant distance from A. Only points in parts (g) and (h) break co-linearity, but maintain the constant distance. That is, they maintain one property but not the other. Breaking one property in this way causes deliberate cognitive conflict and is designed to motivate the student to think deeper about the mathematical properties being considered. These issues are considered further by, for example, Watson and Mason (2006). By this stage, a sensitive student should have sufficient 'practice'. They may have asked question 4 for themselves. It certainly begs to be resolved.

This technique of structuring the examination of mathematical properties of objects can be formalized. We begin with particular mathematical objects, e.g. above we considered points in the plane, and now we turn our attention to polynomials. These objects have mathematical properties, and the structured sequence of questions asks students to create examples of these objects which satisfy a list of criteria which are progressively restrictive. For example, given criteria A, B, and C:

1. Give an example of x satisfying A.
   Give an example of a cubic polynomial.

2. Give an example of x satisfying A and B.
   Give an example of a cubic polynomial with at least one stationary point.[1]

3. Give an example of x satisfying A and B and C.
   Give an example of a cubic polynomial with at least one stationary point and three real roots.

4. Give an example of x satisfying A and B but not C.
   Give an example of a cubic polynomial with at least one stationary point but with fewer than three real roots.

5. Give an example of x satisfying A but not B and not C.
   Give an example of a cubic polynomial without a stationary point.

Note that once a cubic has been sought satisfying all properties, the list is reversed and the final condition negated to demonstrate the boundaries between concepts. Structuring questions in this way has been examined through the work of, for example, Watson and Mason (2002b), Mason and Pimm (1984) and Mason and Watson (2001). All these questions ask for examples of mathematical objects. The role that examples play in teaching and learning of mathematics has received much attention; for example, Michener (1978), Polya (1973), Watson and Mason (2002b); and in the development of the subject as a whole, Gelbaum and Olmsted (1964), and Lakatos (1976). Research (Dahlberg and Housman, 1997, p. 293) concluded that *'The generation of and reflection upon examples provided powerful stimuli for eliciting learning events'*, but that students show a reluctance

---

[1]. Given a polynomial $p(x)$, a *stationary point* is a point x^* at which the derivative is zero, i.e. $p'(x^*) = 0$.

to generate their own examples (Moore, 1994). Such questions can be difficult and time-consuming to assess by hand, but in many situations CAA can help. See, e.g., Figure 1.2 or Figure 7.3.

These questions appear to be 'open-ended', since there is certainly scope for many different answers to each part. But in fact there is a clear purpose and end-point.

> A problem in which the end is 'open' is a recipe for chaos, and conveys a false image of mathematics (akin to the popular image of sociology). [. . .] What we should be trying to encourage in the first instance is neither the property of being 'open-ended', nor that of being 'open-beginninged', but rather that of being *open-middled*. (Gardiner, 2006)

The teacher can then work through a carefully planned worked solution to bring together the essential ideas in full generality. Naturally, the problem must be carefully chosen, so it is interesting and accessible. Students must have the necessary ideas and techniques available from recent classes. What this approach and the examples in this chapter have in common is the belief in the importance of carefully designed questions.

4.5 Types of mathematics question

Chapter 3 examined traditional word problems, and earlier in this chapter we looked at routine techniques. The purpose of this section is to consider and classify other types of mathematical tasks. Since assessment is such a powerful tool for influencing students' choice of learning style, e.g. Gibbs (1999) and Gibbs and Simpson (2004), a discussion of different tasks is, it is hoped, useful in helping informed choices in assessment design and CAA.

Perhaps the best known classification of tasks is Bloom's taxonomy; see Bloom *et al.* (1956). This was refined in a mathematics specific way by Smith *et al.* (1996) in which examination questions are classified into the eight categories shown in Table 4.1. The three groups broadly show an increase in 'level' of the skills. Of course, all such classification schemes are fraught: no attempt is made to classify difficulty. Smith proposed, rather tongue-in-cheek, that *"easy" = questions I can do and "difficult" = questions I can't do*. Similarly, such classifications are entirely context dependent; 5. *Application in new situations*

Table 4.1: The mathematical question taxonomy of Smith *et al.* (1996).

Group A	Group B	Group C
1. Recall factual knowledge	4. Information transfer	6. Justifying and interpreting
2. Comprehension	5. Application in new situations	7. Implications, conjectures and comparisons
3. Routine use of procedures		8. Evaluation

Table 4.2: Pointon's Pointon and Sangwin (2003) question classification scheme.

---
1. Factual recall

2. Carry out a routine calculation or algorithm

3. Classify some mathematical object

4. Interpret situation or answer

5. Proof, show, justify — (general argument)

6. Extend a concept

7. Construct example/instance

8. Criticize a fallacy

---

by definition may only occur once in each mathematical context. We have already acknowledged this as tension 1. Furthermore, we might not be assessing what we intend to assess: a proof requiring only verbatim transcription or minor modification may only test factual recall.

Independent of Smith *et al.* (1996) the question classes shown in Table 4.2 were identified by Pointon and Sangwin (2003), who investigated how hand-held CAS might be used to complete undergraduate mathematics coursework tasks. Their approach was to devise a question classification scheme based on the nature of the task, and then discuss how a hand-held CAS might be used to answer typical questions from each group. These two classification schemes show remarkable similarities, and the majority of questions fit within them.

The remainder of this section provides a brief explanation of the classification scheme outlined in Table 4.2 with some illustrative examples. For more information see Pointon and Sangwin (2003).

1. Factual recall. Almost all questions require some factual recall. However, on occasion questions may require only the recall, usually verbatim, of some knowledge.

2. Carry out a routine calculation or algorithm. Such questions may be complex and involved. They may require the recall of some facts at university level, such as the details of the algorithm itself. They certainly presuppose a certain level of algebraic fluency. Often such tasks may be performed by a computer algebra system.

3. Classify some mathematical object. There are essentially two parts to these questions. Firstly, to recall the definitions, which may be posed as a separate task in its own right. Secondly, to perform any algebraic tasks or provide justification to show the specific object satisfies this. Often this is a question of the form 'Show that ... is a ...'. Furthermore, correctly applying a theorem, for example, tacitly assumes that the case to which the theorem is being applied satisfies the hypotheses. This latter example is a case of object classification.

▼ **Example question 9**

Show that $(\{3, 6, 9, 12\}, \times \bmod 15)$ is a group, and identify the identity element.

Clearly there will be some factual recall, and possibly some routine calculation. The essential distinction between factual recall, use of routine procedures, and classification of objects is the presence of a specific object. The distinction between the latter is typified by the two questions:

> 'Find a solution to the differential equation . . .'
> and
> 'Show that . . . is a solution to the differential equation . . .'

4. Interpret situation or answer. In this case a problem is posed with reference to a physical situation and requires modelling prior to the application of routine procedures. Typically the solution will require interpretation from the mathematical model to the terms in which the question was posed. The student must communicate their solution in the terms used by the question.

5. Prove, show, justify (general argument). Such a question requires a general argument involving abstract or general objects rather than specific examples. In posing such questions, phrases such as 'prove', 'show', 'with justification' are often used. Although used interchangeably, the solution to a 'prove' question typically requires formal definitions, logic, and proof structures such as a contradiction, induction, and exhaustive cases. The solution to a 'show' question would, in general, require an algebraic demonstration of a general but routine result.

▼ **Example question 10**

Show for all $\mu \in \mathbb{R}$ the matrix $\begin{bmatrix} 1 & \mu \\ 0 & 1 \end{bmatrix}$ has repeated eigenvalue 1.

This is a general argument in the sense that the μ is an arbitrary parameter, but there is little in the way of logical structure in the simplest correct solution. Care needs to be exercised when 'with justification' is encountered, as this may be code for 'we want to see your working' in an otherwise routine calculation.

Either direct factual recall and verbatim transcription of a proof, or adaptation of a known proof as template for the task in hand, is not uncommon. Recall Tension 1 in Section 3.5.

6. Extend a concept. Students are asked to evaluate previously acquired knowledge in a new situation. Such a classification is highly context dependent, temporary, and thus fraught. There may be an overlap with other classifications, such as classification of some object. Nevertheless it is a useful distinction, particularly for the course designer.

▼ **Example question 11**

Show that the set of polynomials in x is a vector space over the real numbers. Show that the process of formal differentiation defines a linear transformation on this vector space. [In the context of a level 1 vector spaces course in which a limited number of examples of finite-dimensional vector spaces have been introduced.]

7. *Construct example/instance.* These questions require the student to provide an object satisfying certain mathematical properties.

▼ **Example question 12**

Find a cubic polynomial which is (i) a bijection from $\mathbb{R}$ to $\mathbb{R}$ and (ii) passes through $(-1, 0)$ and $(1, 1)$.

Research, such as Dahlberg and Housman (1997) and Watson and Mason (2002b), has already shown that one very effective way to learn a concept is to generate examples of objects. Asking students to generate their own examples appears to shift their attention from a local focus on the steps of a particular procedure to a more global awareness of the criteria which a correct example satisfies. A detailed discussion of these issues is given by Mason and Klymchuk (2009). Typically there will be many correct solutions to such problems, and automating such questions has been discussed at length by Pointon and Sangwin (2003), Sangwin (2003b), and Sangwin (2005).

8. *Criticize a fallacy.* Finding mistakes in supposed proofs or criticizing reasoning is classified separately. Such questions require higher level skills and a good understanding of the subject matter, see Example question 1, Maxwell (1959) and Northrop (1945).

4.6 Embedding CAA into general teaching

There are other ways to alert students to structure in mathematics by using randomly generated questions in CAA, and also with which to embed CAA within more general teaching. We shall illustrate this technique with an example. It would be possible to do this in class, or with traditional teaching, but here the randomization and immediate feedback both play a crucial role.

Students begin by completing the following CAA questions. There are four separate 'parts', but they all relate to a single cubic and hence they are multiple parts of a single CAA item. Let $a < b < c$ be (reasonable) non-zero integers with $c-b$ even. These are the random numbers, and each student will see one version of this, substituted in the template below.

1. Find an expression representing a cubic polynomial $p(x)$ with $p(0) = 1$ and $p(a) = p(b) = p(c) = 0$.
2. Let $d := \frac{c+b}{2}$. Find the equation of the tangent line to p at $x = d$.

3. Find the intersection of the tangent line with the x-axis.
4. Find all intersections of the tangent line with p.

You are strongly encouraged to do this for yourself before reading on; e.g., try this with $a = -1, b = 1, c = 3$.

The answers to these questions then form part of the following teaching sequence.

1. Each student works on a different, single set of questions 1–4 of their own. In this case it is a task in algebra and calculus, with CAA support.
2. Students compare their examples, working, and results. They all get the 'same answer', which suggests a curious invariant. In particular, they should notice that the intersection points in parts 3 and 4 are both the other root, i.e. $(a, 0)$.
3. Once this is spotted, they might be encouraged to create a dynamic worksheet, so they can move the parameters for themselves. Here we cross from 'nice integers' to continuous real numbers. Also, students move from an algebraic/symbolic register to a graphic/dynamic one. Mixing representations in this way may be cognitively helpful. Such a worksheet, in GeoGebra, is shown in Figure 4.2. Points A, B, and C can be dragged, and the cubic and tangent lines are updated dynamically. Alternatively, a teacher could demonstrate the invariant with such software in class.
4. Then, as a separate teacher-directed task, they have to prove this conjecture in general, again using algebra and calculus.

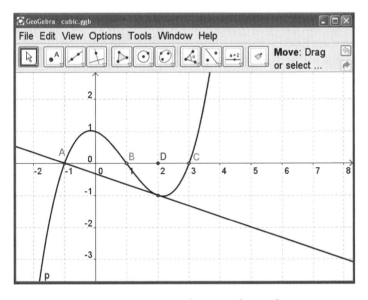

Figure 4.2: Dynamic mathematics with GeoGebra.

5 There are other lines of enquiry. Does $p(0) = 1$ matter? Can a similar result be found when a cubic has only one real root? (Hint: start with the algebra of lines tangent to the cubic and interpret the answer geometrically.)

This sequence is very fragile indeed to *every student answering questions 1–4 correctly*. Without this there is no invariant for them to spot. CAA can play a crucial role in this teaching scenario by (i) generating the random versions, and (ii) providing immediate feedback to students on their work, ensuring they at least eventually reach the correct answer themselves. This feedback is *private* before students are asked to contribute in public to a whole-class discussion. It would have been possible to tell the students that the line tangent to a cubic at the average of two adjacent real roots intersects the cubic at the other root. Instead, students are being led to a general result, a theorem even, through the use of specific examples which they have worked on themselves.

It is sometimes difficult to see how a particular topic might be reframed into this style. Indeed, it might even be claimed that in some areas of mathematics this approach is impractical. However, there is a tradition of setting out entire university courses using almost nothing but carefully structured problems, with perhaps the occasional additional paragraph of historical information or introducing notational conventions. Even *definitions*, e.g. the standard ϵ/δ approach to continuity of real functions, can be built up as the inevitable consequence of carefully structured exercises. This approach is, however, fragile. Asking students to state generalizations too early can result in muddled sentences, incorrect statements remembered, and general confusion.

> I do not wish to belittle the tremendous improvement in instruction which results from pursuing the inductive method, in contrast to authoritarian procedures. [...] It is recognition of the nonverbal awareness stage in inductive learning that converts the classroom experience into that of actual discovery, the kind of thing that promotes a taste for and a delight in research. (Hendrix, 1961)

There are a number of examples of such extended problem sets. For a first calculus course see Wall (1969), for pure Euclidean geometry, leading to other geometries, see Yates (1949), for links between group theory and geometry see Burn (1987), for number theory see Burn (1996) and for undergraduate real analysis see Burn (2000). A particular form of teaching in this style, in which students are given the responsibility of presenting their solutions to problems to the class, is known as Moore's method; see Parker (2004).

4.7 Conclusion

In this chapter we have considered a mathematical exercise as an example of technique and how *domains of variation* and *ranges of permissable change* help to structure educationally equivalent questions. Invariance of the structure of steps in the worked solution under randomization defines *educational equivalence* of versions of a question. These steps in a worked solution define the level of detail appropriate to the group of students for whom the questions are intended. If we have different 'steps' we probably have different questions. Randomization can also be implemented by selecting from a list of question templates to provide a sample of 'different' questions. We have also considered how questions are sequenced, for practice of techniques and to build conceptual understanding. Lastly, we have examined how some of these features might be captured in computer aided assessment.

5

Notation and syntax

> Examples of the power of a well contrived notation to condense into small space, a meaning which would in ordinary language require several lines or even pages, can hardly have escaped the notice of most of my readers. (Babbage, 1827, p. 331)

This quotation points to the *communicative* power of a good notation, but in addition to this a well-designed notation has the ability to *aid calculation and thought*. One example of this is exponential notation, and the law

$$x^n \times x^m = x^{n+m}. \tag{5.1}$$

The n in x^n indicates the number of terms in the product $x \times x \times \cdots \times x$, and hence (5.1) is true by definition for whole-number indices. But it also suggests the meaning of $x^{\frac{1}{2}}$ or x^{-1}. The meaning is derived from a desire to keep the law (5.1) true. Thus, notation can generate meaning in the name of consistency as well as record the product of thought. Furthermore, by concentrating only on the indices we see that 'multiplication becomes addition' and the logarithm is suggested. However, mathematically, pedagogically, and historically things are not so tidy. Indeed, a substantial shift is required to move from an elementary interpretation of $x^{\frac{n}{m}}$ to a formal definition in analysis; Gardiner (2003).

Effective mathematical notation also helps with *recognition*. By writing mathematical expressions in standard forms they can be recognized and routine techniques applied to them. Communication is closely related to recognition. Indeed, Corless et al. (1996) cite numerous applications of the Lambert W function, defined as the multivalued inverse of the function

$$w \to we^w$$

(just as the complex logarithm is a multivalued inverse of $w \to e^w$). This function has not always been recognized by mathematicians since its first use around 1758, and Corless et al. (1996) conclude as follows. *'Names are important. The Lambert W function has been widely used in many fields, but because of differing notation and the absence of a standard name, awareness of the function was not as high as it should have been.'*

Mathematical notation is the product of a long tradition, and has evolved to take advantage of a rich set of special symbols, together with their relative size and position on a two-dimensional page. However, when typing a mathematical expression using a keyboard we have only a one-dimensional string of symbols taken from a limited alphabet. Translating mathematics into such a format is a fundamental problem.

In this chapter we consider how a student might enter their mathematics into a machine. This is a key step in the assessment process. In Section 3.5 it was stated that *'the outcome of mathematics is a correct chain of reasoning, from agreed hypotheses to a conclusion'*. If we are to automatically assess mathematics we should attempt to look at the whole answer. However to begin we must accept a single mathematical expression as an answer. This will be sufficiently difficult (and interesting) to occupy us in this chapter, and the next will consider how to establish properties of a single expression.

Many users need to interact with a machine mathematically, but here we draw a distinction between professional users and students. Within the 'student' category a distinction is drawn between two *modes of use*: on one hand, problem-solving and calculation, and on the other, assessment. It is in the context of assessment that these issues become especially problematic. Sangwin and Ramsden (2007) have argued that

> whatever the benefits or drawbacks of a non-standard, especially precise syntax in problem-solving and calculation, such a syntax has clear disadvantages if students taking tests are required to use it, precisely because of the risk of failing on a technicality. This, we feel, is unacceptable.

Indeed, syntax presents the most significant barrier to students' successful use of computer aided assessment, particularly when the stakes are high.

Perhaps the most common two ways to interact with a computer are the keyboard and the mouse, or similar pointing device. Voice and handwriting recognition are becoming more widely available, and with them attempts to incorporate mathematics. This chapter reviews all these methods of mathematical input. Of course, while our main focus is CAA, input is a more general issue. First we will consider the history and meaning of mathematical notation.

5.1 An episode in the history of mathematical notation

> Symboles are poor unhandsome (though necessary) scaffolds of demonstration; and ought no more to appear in publique, then the most deformed necessary business which you do in your chambers. (Hobbes, 1656)

We shall consider only one episode in the history of mathematical notation: that of the development of exponential notation and the law now expressed as (5.1). Much more detail is given by Cajori (1928), still an important general historical source on notation. It is necessary to preface this sketch with a caveat: because someone has priority for the first use of a particular notation does not mean that they had any significant influence. For example, Nicolas Chuquet's (1445–1488) work *Triparty en la science des nombres*, written around 1484, contains a number of important and prescient ideas, but it was unpublished until 1880; Chuquet (1880). Similarly, some older notations remained in use long after more modern innovations occurred. Here, William Oughtred's (1574–1660) *Clavis Mathematicae* (Key to Mathematics) is a prime example; see Oughtred (1652) and the commentaries by Stedall (2000, 2002). Furthermore, large geographical distances hampered communication, and the inconsistent use by authors over numbers of years make the history of this topic complex.

Originally, writers used a rhetorical style, writing out mathematics as full prose. One of the motivations for algebraic symbolism is *abbreviation*. This is certainly articulated in the quotation of (Babbage, 1827, p. 330) at the beginning of the chapter. Gradually single letters, e.g. the initial letters of words, were used as abbreviations. William Oughtred, following Viete, took this approach in his work *Clavis*. For example, he used the capital letter C to denote the cube of the unknown. The word for unknown is in Latin *causa*, in Italian *cosa*, and in German *coss*. Hence algebra became known in England as 'the Cossic Art' or Art of Things. Luca Pacioli (1445–1517), for example, wrote *co* in mathematics to refer to the unknown quantity as an abbreviation. Gradually notation developed into the symbolic algebra we know today. Table 5.1 compares mathematical notation, some of which are surprisingly hard to read by modern standards.

Modern exponential notation begins, really, with René Descartes' (1596–1650) *Géométry* published in 1637; Descartes (1952). He was the first writer to use recognizably modern notation such as $5a^4$. Other writers had come close. In 1634, Pierre Hérigone (1580–1643) wrote 5a4 placing coefficients before the variable and exponents after, but on the same line; see Hérigone (1634). In 1636 James Hume brought out a new version of Viete's work (see Viete (1636)) in which he wrote $5a^{iv}$, using Roman numerals for exponents.

However, Descartes preferred aa to a^2. Such repetition to denote powers was common in the seventeenth century: Thomas Harriot, John Newton, and John Collins all made use of it (Cajori, 1928, §307). However, it was Descartes' notation which gradually gained

Table 5.1: The comparison of mathematical notation by Babbage (1830).

Author	Example
Pacioli	1*cu. m.* 6*ce. p.* 11 *co. equale* 6n^i
Bombelli	1$\overset{3}{\smile}$ .m. 6$\overset{2}{\smile}$ p. 11 $\overset{1}{\smile}$ equale 6
Stevinus	1③ – 6② + 11① : *eguale* 6
Viete	1C – 6Q + 11N *egal* 6
Harriot	1.*aaa* – 6.*aa* + 11.*a* = 6
Modern	$x^3 - 6x^2 + 11x = 6$

ground. For example, John Wallis (1616–1703) used it in his *Arithmetica Infinitorum* of 1655, despite being taught with Oughtred's notation; see Wallis (2004).

Descartes used this notation only for positive integer powers. Examples of negative and fractional powers are cited by (Cajori, 1928, §308), notably by Wallis, but it was Newton who made what is perhaps the first confident statement on this subject.

> Since algebraists write a^2, a^3, a^4, etc., for $aa, aaa, aaaa$, etc., so I write $a^{\frac{1}{2}}, a^{\frac{3}{2}}, a^{\frac{5}{2}}$, for $\sqrt{a}$, $\sqrt{a^3}, \sqrt{a^5}$; and I write a^{-1}, a^{-2}, a^{-3}, etc., for $\frac{1}{a}, \frac{1}{aa}, \frac{1}{aaa}$, etc. (Horsley, 1782, p. 215)

He then uses this notation in his binomial formula for $(x+y)^{\frac{m}{n}}$. It was Leonhard Euler (1707–1783) who first introduced imaginary powers, on 18 October 1740 (Cajori, 1928, §309), allowing him ultimately to write his eponymous formula in a recognizably modern form.

> From these equations we understand how complex exponentials can be expressed by real sines and cosines, since $e^{iv} = \cos v + i \sin v$ and $e^{-iv} = \cos v - i \sin v$. (Euler, 1988, §138)

Note, however, that in the original of 1747, Euler had not yet adopted i as notation for $\sqrt{-1}$, so the formula originally appeared as $e^{+v\sqrt{-1}} = \cos .v + \sqrt{-1} \sin .v$. Euler goes even further in making use of the exponential notation, the meaning of which is not entirely clear.

> We may even apply this method to equations which go on to infinity. The following will furnish an example:
>
> $$x^\infty = x^{\infty-1} + x^{\infty-2} + x^{\infty-3} + x^{\infty-4} +, \&c.$$

(Euler, 2006, §799)

From the mid eighteenth century notation for exponents appears thoroughly modern. It is important to keep in mind the period of time over which these developments take place. Almost 100 years elapsed between Descartes and Euler, during which many competing notations were used by professional mathematicians for exponential notation alone—a notation we take almost for granted.

5.2 The importance of notational conventions

In Chapter 3 we discussed tensions, and in particular we highlighted the problem of recognition. Using notational conventions allows a problem to be recognized as one to which a standard technique can be applied. This is a key step in problem-solving at all levels. By using notation in conventional ways, problems become much easier to solve: it is much easier to recognize abstract things when written in standard forms. The consistent and confident use of notation in this way is an important part of mathematical practice.

However, conventions should be distinguished from the logical consequences of assumptions. If a student chooses not to follow conventions and yet is clear and correct in their reasoning then the assessment criteria should accommodate this.

The first step to solving any problem is to identify the relevant information, either from the problem itself or from cultural knowledge, e.g. metrological units and physical laws. This is abstracted into a mathematical formulation, for example as algebraic equations, geometric relationships, or differential equations. During the process, units and dimensional consistency provide useful checks and allow certain types of absurdities to be spotted.

When abstracting a given problem into a mathematical formulation it is often helpful to understand and adopt appropriate conventions. For example, in algebra letters towards the beginning of the alphabet, a, b, c, are used to denote *known but as yet unspecified numbers*, while those at the end, x, y, z, are generally used for *unknown numbers*, which represent the solution to the problem in hand. Then equations and algebraic expressions have *standard forms*, such as those with ordered gathered terms. Compare

$$x^2 + 3x - 1 = 0, \quad \text{vs} \quad x + x^2 = 1 - 2x.$$

We shall discuss forms of elementary algebraic expressions more fully in Section 6.7.

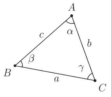

Geometry also has the following conventions:

1. *points* are upper case Roman letters, A, B, C etc.;
2. *lines* or *segments* or *curves* are lower case Roman letters, a, b, c etc.;
3. *angles* are Greek letters, α, β, γ etc.

Notice that in a triangle ABC the side opposite, a, the point A, and the angle α all correspond. Then it is immediately clear that 'side a in ABC' refers to the side opposite A.

We also regularly make contextual assumptions about particular letters in the alphabet: f is a function, t is time, and n, m are integers. These do reduce the cognitive load and aid recognition. So, while a CAS might treat letters as abstract symbols without context or assumptions, in mainstream usage certain letters *do* fulfill traditional roles.

> The advantage of selecting in our signs, those which have some resemblance to, or which from some circumstance are associated in the mind with the thing signified, has scarcely been stated with sufficient force: the fatigue, from which such an arrangement saves the

reader, is very advantageous to the more complete devotion of his attention to the subject examined. (Babbage, 1827, p. 370)

For example, Babbage suggests the use of t to denote a time variable. He also recommends signs where the meaning is closely associated with the shape such as $=$, $<$ and $\leq$. Contemporary thought agrees; for example:

> In choosing infix symbols, there is a simple principle that really helps our ability to calculate: we should choose symmetric symbols for symmetric operators, and asymmetric symbols for asymmetric operators, and choose the reverse of an asymmetric symbol for the reverse operator. The benefit is that a lot of laws become visual: we can write an expression backwards and get an equivalent expression. For example, $x + y < z$ is equivalent to $z > y + x$. By this principle, the arithmetic symbols $+ \times < > =$ are well chosen but $-$ and $\neq$ are not. (Hehner, 2004)

Ultimately, and somewhat unusually, Babbage sets out to discuss his *principles for mathematical notations* given in Table 5.2.

It is interesting for us to reflect on these today, as they are by no means universally accepted; for example, (B5) is used for trigonometric functions in the English speaking world but not in other European traditions, and is nowhere used for exponential functions.

Table 5.2: Principles for mathematical notation. From Babbage (1830).

(B1) All notation should be as simple as the nature of the operations to be indicated will admit.

(B2) We must adhere to one notation for one thing.

(B3) Do not multiply the number of signs without necessity.

(B4) When it is required to express new relations that are analogous to others for which signs are already contrived, we should employ a notation as nearly allied to those signs as we conveniently can.

(B5) Whenever we wish to denote the inverse of any operation, we must use the same characteristic with the index -1.

(B6) Every operation ought to be capable of indicating a law.

(B7) It is better to make any expression an apparent function of n, than let it consist of operations n times repeated.

(B8) All notations should be so contrived as to have parts being capable of being employed separately.

(B9) All letters that denote quantity should be printed in italics, but all those which indicate operations, should be printed in a Roman character.

(B10) Every functional characteristic affects all symbols which follow it, just as if they constituted one letter.

(B11) Parentheses may be omitted, if it can be done without introducing ambiguity.

5.2 THE IMPORTANCE OF NOTATIONAL CONVENTIONS | 59

In addition to these, and for the purposes of devising a typed linear syntax for CAA, Sangwin and Ramsden (2007) augmented this list with the following:

(P1) *Informal linear syntax should correspond with printed text and written mathematics.*
Students can rightly expect us to be consistent in the way mathematics is expressed, as far as is reasonable given the constraints of a one-dimensional input mechanism.

(P2) *Informal linear syntax should not obstruct learning the strict syntax of a CAS.*
There should be nothing to unlearn at a later stage.

However, as we shall see, there are some serious problems with traditional notation which make fulfilling these ambitions impossible in a syntax for CAA.

So far in this chapter we have made some strong statements about the importance of notation. In case you are in any doubt we will now consider a *change* in the notation used for the logarithm function. Surely, if notation is unimportant then change should be trivial. The proposal, made by Brown (1974), was inspired by computer notation for exponentiation a^b as $a \uparrow b$, or perhaps more commonly now, a^b. The proposal is for the logarithm of b to the base a, sometimes written as $\log_a(b)$, to be written $a \downarrow b$. We might choose to write this as a_b, and it would be perfectly possible to type a_b into a machine as a_b. Then the natural logarithm becomes simply e_x. In this new notation how do we write the laws of logarithms? How do we solve traditional problems involving logarithms?

Formally at least, $a^{a_x} = a_{a^x} = x$. To solve $a^x = b$, we 'operate' with '$a \downarrow$' to obtain

$$a^x = b, \quad \Leftrightarrow \quad a_{a^x} = a_b, \quad \Leftrightarrow \quad x = a_b.$$

Furthermore,

$$a_{b \times c} = a_b + a_c, \quad a_{b^c} = (a_b) \times c, \quad (a_b) \times (b_a) = 1, \quad (a_b) = (a_c) \times (c_b).$$

It has been more than 35 years since the publication of Brown (1974) and yet, sadly perhaps, this innovation appears to have made no impact!

Notice how so many of these laws now take advantage of the positional features of the notation. These symbolic patterns have a certain aesthetic appeal, which their traditional counterparts using $\log_a(b)$ do not. To investigate such ideas, Kirshner and Awtry (2004) introduced the concept of *visual salience*.

> The quality of visual salience is easy to recognize but difficult to define. Visually salient rules have a visual coherence that makes the left- and right-hand sides of the equations appear naturally related to each other.

They go on to analyse students' errors in terms of visually salient rules. They found that visually salient rules were both easiest to learn and most likely to be over-generalized. Traditional artificial intelligence models students' errors as 'buggy rules' (see Section 6.10). In the AI model students are using declarative knowledge, albeit misslearned. However, Kirshner and Awtry (2004) argue instead that students are operating only at the surface level of visual pattern matching, and in terms of 'animated sequences' of moving symbols as the eye tracks across the equality sign. Related to this Kirshner (1989) found that *'for some students the surface features of ordinary notation provide a necessary cue to successful syntax decisions'*. So while the power of notation lies partly in its visual appeal, students still need to ensure that they understand the limitations so that they are not simply symbol-pushing in a mindless way.

5.3 Ambiguities and inconsistencies in notation

Although mathematics is a precise discipline where clarity is valued, traditional mathematical notation contains some inconsistencies and ambiguities. Consider the following expressions:

$$3\frac{2}{5}, \quad 3(2+5) \quad \text{and} \quad x(t+1).$$

In the first we have three and two fifths. In this expression the juxtaposition of 3 with $\frac{2}{5}$ is an implied addition. In the second expression the juxtaposition is taken to mean a multiplication. The third expression is ambiguous since it could be a multiplication, but it could equally be an evaluation of the function x at $t+1$. Only the context would decide. Another ambiguity arises in combinatorics and probability, where multiplication of integers is sometimes denoted using a centre dot, so that

$$\frac{4 \cdot 3}{2 \cdot 1} = 6,$$

rather than interpreting $\cdot$ as a decimal separator.

Next we turn our attention to the inverse trigonometric functions. Corresponding with (B5), contemporary usage in the United Kingdom is to write the inverse of sine as $\sin^{-1}(x)$. However, the square of the value of sine is often written $\sin^2(x)$. Consistency would require us to have exponents operating on the value of the function, so that

$$\sin^2(x) = (\sin(x))^2, \text{ and } \sin^{-1}(x) = \frac{1}{\sin(x)},$$

or for exponents to denote composition giving

$$\sin^2(x) = \sin(\sin(x)) \text{ and the inverse as } \sin^{-1}(x).$$

Babbage preferred the second of these two interpretations. Also, given (B11), he omits parentheses writing only sin x for $\sin(x)$. Then we can apply laws similar to (5.1) and write

$$\sin^{-1} \sin x = \sin^{(1-1)} x = \sin^0 x = x.$$

Going even further, these considerations led him to his 'calculus of notations': Babbage (1821). The meaning he ascribes to $\sin^{\frac{1}{2}}$ is the function which, when composed with itself, gives sin. We do not comment here on the circumstances under which this is mathematically legitimate.

5.4 Notation and machines: syntax

Our goal is to enable students to input their answers into a machine for the purposes of CAA. Given these ambiguities, even at the most elementary level, we have some difficulties to resolve. It is clear that we cannot simply implement a 'what you see is what you get' approach, replicating traditional notation in computer form.

The most basic interaction with a computer is to type at a keyboard. It seems likely that keyboards will persist for some years, and so mathematical expressions will be encoded into a string of symbols. The first problem is that of assigning definite meanings to individual symbols. According to (Cajori, 1928, (I), p. 165) the symbol = was probably introduced in 1557 by Recorde (1557) as a synonym for the phrase 'is equal to'. There are at least four very common senses in which this symbol is currently used.

1. To denote an equation yet to be solved ($x^2 + 1 = 0$).
2. Assignment of a value to a variable ($x = 1$).
3. Definition of a function ($f(x) = x^2$).
4. As a binary Boolean infix operator, i.e. a function returning either TRUE or FALSE.

Further uses in computer algebra were identified by Davenport (2002) and Bradford et al. (2009). Using the same symbol for each of these senses certainly violates Babbage's (B1), and the usage is not symmetric in 2 and 3. Further, we might view assignment of a value as a special kind of function. Indeed, the CAS Mathematica distinguishes between *immediate assignment*, using =, which happens when the variable or function is defined, and *delayed assignment* := which happens when the variable or function is called. So here, the appropriate assignment for variables is usually immediate and functions delayed. In similar ways, these different senses of 'is equal to' are encoded in the typed syntax of different CAS. Table 5.3, from Sangwin and Ramsden (2007), gives us a comparison of the syntax choices made by Axiom, Derive, Maple, Mathematica, and Maxima.

Table 5.3: Expressing different forms of equality.

CAS	Assignment	Equation	Function definition	Boolean infix
Axiom	:=	=	==	=
Derive	:=	=	:=	=
Maple	:=	=	:=	=
Mathematica	= (or :=)	==	:= (or =)	==
Maxima	:	=	:=	=

Next we have to link symbols into one dimensional strings. We cannot take advantage of relative size and position on the page, as we would in free handwriting. For example:

$$3(x^2 + 1)\sin(x)$$

might be typed in as `3*(x^2+1)*sin[x]`. Notice the explicit multiplications which are implicit in the expression, and the difference in using parentheses for grouping and square brackets for function arguments. Almost all mainstream CAS use such a *linear direct algebraic logic syntax*, with infix binary operators for arithmetic. This corresponds closely to traditional written mathematics, e.g. 'two times x' is expressed as `2*x`.

Briefly contrast this with the so-called Reverse Polish Notation (RPN). In RPN, operators follow their operands; for example, `2 x *`. When we have many operations these follow in order, so $2x + 1$ would be entered as `x 2 * 1 +`. One advantage is that parentheses are no longer needed. $2(x + 1)$, which requires parentheses in traditional notation, is written as `1 x + 2 *` in RPN. Another advantage is that interpreters of RPN are often easy to implement and are very fast. They also connect directly with the tree structure which is implicit in the rules of precedence.

The work of Sangwin and Ramsden (2007) examined the CAS's 'basic' linear syntax, as strings of keyboard characters. In particular they reviewed the syntax of Derive, Maple, Mathematica, Maxima, and Axiom, all of which have been used in online CAA. Some results are illustrated in Tables 5.3, 5.4, and 5.5. There are clearly many differences, and while this appears to be rather trivial, actually the student being assessed needs to understand these conventions to successfully communicate their mathematics.

Strings of symbols, representing numbers only, were typed into the CAS, and Table 5.4 shows how these were interpreted. Notice that when digits are grouped and separated with a space, e.g. `23 000`, only Mathematica and Derive accept this as valid, but with the interpretation of implied multiplication. In all systems, except Derive and Mathematica, the string `2.5e-2` is interpreted as scientific notation for the floating point number 0.025. Mathematica and Derive are more explicit, requiring, e.g., `2.5*10^-2`.

The unary minus is also interesting. In common with many calculators and textbooks, when a unary minus precedes a number all CAS interpret `-4^2` = $-(4^2)$ = -16. It would

Table 5.4: Entry of numbers.

CAS	23 000	2.6e-2	4^-2	-9^1/2	x+-2	x*-2
Axiom	ERROR	0.026	$\frac{1}{16}$	$-\frac{9}{2}$	$x - 2$	$-2x$
Derive	0	$\frac{13e}{5} - 2$	$\frac{1}{16}$	$-\frac{9}{2}$	$x - 2$	$-2x$
Maple	ERROR	0.026	ERROR	$\frac{-9}{2}$	ERROR	ERROR
Mathematica	0	$-2 + 2.6e$	$\frac{1}{16}$	$-\frac{9}{2}$	$-2 + x$	$-2x$
Maxima	ERROR	0.026	$\frac{1}{16}$	$-\frac{9}{2}$	$x - 2$	$-2x$

Table 5.5: Entry of special symbols.

CAS	π	e	$\sqrt{-1}$	∞
Axiom	%pi	%e	%i	%plusInfinity
Derive	pi	#e	#i	inf
Maple	Pi	exp(1)	I	infty
Mathematica	Pi	E	I	Infinity
Maxima	%pi	%e	%i	Inf

be reasonable to expect the unary minus to bind more tightly than any other operator, to signify that -4 is a single entity, the number minus four. Then $-4\verb|^|2 = (-4)^2 = 16$.

Entry of constants π, e, $\sqrt{-1}$ and ∞ are shown in Table 5.5. Here ∞ is the positive real infinity, and CAS usually differentiate between this, negative real infinity, and complex infinity. None of the systems used j in place of i to denote $\sqrt{-1}$. Note that in all the CAS, both i and e are interpreted as arbitrary variables, not mathematical constants. Hence in e^x the e is simply an arbitrary and undefined variable. Similarly, in Maple, pi and Pi are different, with the former simply being the variable denoted by the Greek letter. However, both are displayed exactly as 'π'. This is not the case with the Greek letter gamma which is interpreted by Maple, not as a variable, but as Euler's constant, $\gamma \approx 0.5772$.

The arithmetic binary operations $+$, $-$, and $/$ were identical in all systems. Exponentiation is denoted using the symbol ^ in all systems, with Axiom also using $**$. All but Derive and Mathematica were strict in requiring an explicit multiplication sign $*$. Derive does allow implied multiplication under some circumstances; for example, it interprets 2x as 2*x, and we note that Mathematica supports two typed syntaxes: InputForm and TraditionalForm. Further details are given in Sangwin and Ramsden (2007).

Nouns and verbs

The traditional use of a CAS is as a 'super-calculator'. It is therefore natural that the CAS will seek to simplify expressions as far as possible. In particular if a user types in an expression such as $x + 5x + 1$ the CAS will *perform the addition* to write this as $6x + 1$. Here the user has implicitly asked for the $+$ symbol to be a *verb*. That is to say, the user wants the addition to be performed. However, there are plenty of situations when the user wants only to *represent* addition and not *perform addition*.

When a user wishes to indicate the representation of some operation, we say they use the *noun* form of the function, and when they want to perform the operation they are making use of a *verb*. This is not so unusual as you might suppose. For example, consider the differential equation

$$\frac{d^2x}{dt^2} = -x.$$

In order to enter the equation into a machine, we must represent the second derivative of x with respect to t. However, we do not want the machine to perform the differentiation in response. We are not issuing a command to actually differentiate the symbol x at this movement. When typing the command such as `diff(x,t,t)` (or the equivalent) into most CAS we get 0. It may be permissible to indicate that x is a function of t with some syntax such as `diff(x(t),t,t)`, but this confuses the need for the noun form of differentiation with the evaluation of a function, or rather its second derivative, at a particular point (in time) t.

We shall consider noun and verb forms in much more detail in the next chapter. Suffice to say now, for the purposes of CAA, we must assume that all the operations entered by the student are interpreted as nouns, right down to the level of leaving expressions such as 1 + 1 unevaluated. Really, we should have some consistent mechanism for indicating that *any* function within the CAS is a noun within an expression, and for converting operators between noun and verb forms.

These ideas can be found in implementations. Maple, for example, uses an initial capital letter to denote a noun form which is not simplified. So, the differential would be entered as `Diff(x,t,t)`. In Maxima, using the prefix ' prevents evaluation. To obtain the noun form of the elementary arithmetic operations we have to use the prefix form; that is to say, 1 + 2 is typed in as `"+"(1,2)`. The noun version is then obtained as `'"+"(1,2)`. This is hardly a natural syntax for students, but at least the CAS is capable of being able to represent such an expression. We shall return to this issue in Chapter 6.

Functions

Perhaps the most interesting differences are in the way that CAS represent functions. Some CAS allow the manipulation of whole functions, or rather the names of functions, as legitimate objects. Maple's D operator acts on a function; for example, `D(sin)` simplifies to cos. Hence, we are dealing with an unevaluated function. The expression $\sin^2$ when applied to x is interpreted as $\sin(x)^2$, not as a composition, and the input `sin^(-1)` is interpreted in a consistent way as $\frac{1}{\sin}$, rather than the inverse. Axiom and Maxima rejected all attempts to exponentiate functions without arguments.

It is more usual with a CAS to deal with a function evaluated at a point, e.g. $\sin(t)$ even if t has not been assigned a specific value. The syntax for such functions shows significant systematic differences. Mathematica's input form uses the function name followed by the argument enclosed in brackets (for example `f[x]`), and parentheses are reserved for grouping terms. Other CAS use parentheses instead: `f(x)`. The use of square brackets avoids the most obvious ambiguity, leaving juxtaposition free to mean implied multiplication, but of course square brackets do not correspond with written tradition.

For the trigonometrical functions all CAS use `sin`, `cos`, `tan`, with Mathematica using a capital initial letter. In all but Derive, which offers a choice, the system enforces *radian angular measure*. For inverse trigonometrical functions Maple uses the pattern following `arcsin`, with `Arcsin` treated as a noun. Mathematica uses `ArcSin`, Axiom, Maxima, and Derive use `asin`, and so on.

In addition, both Derive and Axiom accept a space to signify function application. For example, `sin x` is interpreted as $\sin(x)$ with parentheses required only to group terms, e.g. `cos (n*pi)`. Here the space is optional. Using a space in this way reduces the number of symbols. Further, `sin sin x` is interpreted as a composition. Derive is unique in accepting `sin^2(x)` as $(\sin x)^2$, and is consistent in interpreting `sin^-1(x)` as $\frac{1}{\sin x}$. One quirk unique to Maple is the interpretation of `2(x+1)` as the application of the constant function 2 to the argument $(x + 1)$, which results in the value 2. Hence `sin^2(x)` is the sine function raised to the power of `2(x)`, which is the constant function 2 and argument x. This is interpreted as $\sin^2$, which is a function, rather than the result of applying a function to an argument.

All systems implement the exponential function as `exp`, with Mathematica using its consistent variation `Exp[x]`, and `log(x)` refers to the natural logarithm.

5.5 Other issues

We have seen significant differences in CAS syntaxes, even at the most elementary level. Such differences become compounded: choices at one level affect those at the next. Since computer systems are unforgiving to even minor syntax errors, these apparently minute variations really do matter. While students are likely to use only one CAS regularly, similar disconnects occur with calculators, other software, and with the syntax for programming languages. This section mentions a number of important and related issues: a design decision needs to be made on each of these.

The first is to acknowledge cultural differences in notation, see Libbrecht (2010). For example, in the United Kingdom, Ireland, and their former colonies the decimals separator is a full stop. However, most of Europe uses a comma. The extent to which digits are grouped to ease the reading of longer numbers varies from country to country, as do the symbols used to separate the groups. A full stop, comma, space, or apostrophe are all used for this purpose.

Denoting intervals of the real line also reflects cultural differences. For 'half open' intervals $0 < x \leq 1$ the Anglo-saxon way is $(0, 1]$, whereas the French way is $]0, 1]$. Notice that this notation for intervals is an obvious overloading of the notation for coordinates (x, y) or lists $[x, y]$. In all the CAS, an expression such as `(1,2]` was rejected as syntactically invalid.

What should we do about symbols which do not appear on the keyboard? Greek letters can be typed, as in `pi`, but what about the integral $\int$, or plus or minus, $\pm$?

There are similar differences between CAS in the syntax for Boolean functions (*verb* forms) for logical NOT, AND, OR, and so on, and corresponding *noun* forms used as connectives, which we do not detail here.

Inequalities can easily use their keyboard symbols, with non-strict inequalities using `<=` or `>=`. But, only Mathematica accepts chains of inequalities, such as $1 < x < 2$, interpreting this as a list of inequalities, all of which must hold.

5.6 The use of the AiM system by students

The work of Sangwin and Ramsden (2007) also examined students' attempts at using the AiM computer aided assessment system, see Section 7.1, which requires a strict CAS syntax to be used. Errors were numerous and included the following:

1. Missing multiplication signs.
2. Missing parentheses resulting in incorrect grouping.
3. Entry of sets as a comma separated list, i.e. 1,2,3, not {1,2,3}.
4. Where syntax hints are given, they are often forgotten.

The largest source of students' errors was, perhaps unsurprisingly, a missing *.

Many CAA system designers have resolved these problems by implementing a sequence of heuristics to disambiguate expressions. In limited areas this works well, and the decision to opt for an input mechanism with such an 'informal syntax' depends on the particular student group and their needs. Alternatively, there will be students who need to learn exactly the syntax of a CAS in order to communicate with it as a useful tool.

Sangwin and Ramsden (2007) concluded:

> Our research and development work in CAA suggests that the conflicts are so serious that it is *impossible* to implement an informal *syntax*. Instead a more protracted process is necessary during which the student gradually refines their input in the light of feedback from the system.

They go on to say:

> These issues illustrate clearly, in the authors' opinion, that not all problems associated with the interpretation of mathematical expressions in CAA can be solved at the level of the underlying syntax. We argue that some problems can be solved only by ensuring that input systems contain the facility for clarification dialogs or that during CAA authoring it is possible to override default interpretations, or both.

These results are echoed in other studies. For example,

> in scoring real response data, the algorithm was accurate 99.62% of the time, producing incorrect judgements in only 26 of 6,834 cases, all stemming from examinees improperly entering subscripts or text strings into their expressions. (Bennett *et al.*, 1997)

This failure of the algorithm is precisely a failure of students to use the input mechanism effectively.

On the basis of these findings it has been helpful in CAA to separates two kinds of feedback. The first is associated with the *syntax* of the student's answer, the second to its interpretation *semantics*. To provide a consistent interface, feedback based on the syntax should not depend on the context in which a question is taken. So feedback about syntax errors is the same in a formative learning context as a summative test. Feedback based on the semantics of the answer should be at the discretion of the teacher and will depend on the question. This is discussed further in Chapter 7.

5.7 Proof and arguments

So far we have concentrated on how a human can encode a mathematical expression to enable automatic computation, as in a computer algebra system, or as part of automatic assessment. In Section 3.5 we claimed that *'assessment should reflect mathematical practice'*. Accepting this, we should look further than just expressions, i.e. 'answers', and look at the method and the argument used to reach the answer. The challenge becomes how to encode a whole mathematical argument.

Proof assistants are mathematical software which provide support in dealing with mathematical deduction. There is a significant range from automated theorem provers to systems which enable proofs to be checked. Automated theorem provers belong to the research domain. Their goal is to prove new mathematical theorems, although Barendregt and Cohen (2001) comment on their current limitations. *'In some areas of mathematics—such as elementary geometry—there are theorem provers that work well, but in general proof search is not feasible.'* A survey is given by Beeson (2003). In this specialist area the interface is less problematic. In any case, problems are reformulated before they are encoded into the machine; indeed, techniques such as qualifier elimination mean that the problem upon which the software operates may be unrecognizable to a non-expert.

On the other hand, human mathematicians use (and teach) less formal mathematics. Formal mathematics consists of statements built up using only a formal grammar. Informal mathematics is not necessarily imprecise; rather, it is not constrained by a formal grammar. One difficulty of using formal statements is their length. Even a relatively simple mathematical expression encoded as MathML or OpenMath is almost unintelligible without tools to render them in traditional forms. Arguments necessarily consist of chains of expressions. Human mathematicians *compress* concepts into definitions, whereas automatic theorem provers work from axiomatic foundations in an inflexible way. Little seems to have been achieved in reaching consensus on how to express arguments in elementary mathematics, including mainstream algebra, calculus, and geometry. This lack of uniformity extends to traditional practice, rather than technological systems. It is still a design issue.

> We will argue that this is not just a deficiency of the user interface, but that the problem with automated theorem provers is much deeper, it goes to the core of these systems, namely to the formal representation of mathematical knowledge and the reasoning that can be performed with this knowledge. (Kerber and Pollet, 2007)

Even in the research domain the link between computation as in CAS and automated theorem-proving is very weak.

> What is interesting, and surprising to people outside the field, is that the mechanization of logic and the mechanization of computation have proceeded somewhat independently.

We now have computer programs that can carry out very elaborate computations, and these programs are used by mathematicians 'as required'. We also have 'theorem-provers', but for the most part, these two capabilities do not occur in the same program, and these programs do not even communicate usefully. (Beeson, 2003)

We shall return to this theme when we examine examples of CAA in Chapter 8. Very few CAA systems currently attempt to provide an interface to logical argument and proof. Where steps in the working of a problem are provided by students, these are most often expressions filled into pre-provided answer boxes, e.g. as in the CALM system described in Section 8.2. Other systems, such as MathXpert, enable students to specify what action to perform on an expression, from a list of available actions generated by the system. Very few systems, confined to pilot studies with students, have attempted to provide a full interface which combines mathematical expression and logical argument. This remains a challenging problem.

5.8 Equation editors

An alternative to a strict typed syntax is to use a 'drag-and-drop' equation editor. There are many examples, and one is known as DragMath, described in Sangwin (2012). A similar editor was developed for the DIAGNOSYS system described in Section 8.5, and a view of this is given in Figure 8.7. The editor lets users build up mathematical expressions in a traditional two-dimensional way by dragging templates from a palette. A typical screen shot is given in Figure 5.1.

Templates can be assembled to give two dimensional notational features, such as the sum $\sum_{n=1}^{\infty}$ and also the fractions. Note that the fraction on the right is not complete in Figure 5.1. Symbols which do not appear on the keyboard can be selected from a palette, removing the difficulty of knowing the precise syntax, or even being able to verbalize particular symbols, e.g. letters of the Greek alphabet. Equation editors often contain a parser so that strings of

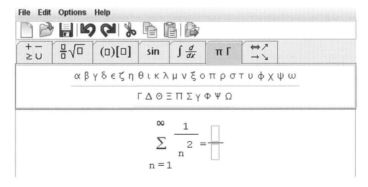

Figure 5.1: The DragMath equation editor.

symbols can be also typed at the keyboard. This helps the experienced user make the most efficient use of the system.

We note that the typesetting systems TEX and LATEX replicate the movable type system of traditional printing in electronic form; Knuth (1979). They do not seek to encode the meaning of mathematical expressions, and from this create a displayed form. Hence, systems such as TEX and input tools such as DragMath replicate traditional notation in electronic form. They have their advantages, but they also capture all the ambiguities and require context.

A number of schemes are used to disambiguate the input. DragMath has an 'unnamed function' application button, to disambiguate function application from multiplication. The user must supply both the function name, and its argument. To the user this appears identical to multiplication. For example, $x(t + 1)$ appears on screen in exactly this form, but internally the data structures are distinguished. Other systems use various heuristics to provide context.

There are some very interesting and subtle design challenges which need to be addressed by such an input mechanism. The most challenging issue is that of *selecting* parts of an expression, either to delete, cut, or paste. The tree structure underlying mathematical expressions provides a natural structure which guides this selection process. Indeed, it is normal to want to select a whole sub-expression. Conversely, it makes little sense to try to select the 'i' in $\sin(x)$. However, with associative binary operations we might only want to select some of the arguments. For example, it is possible we might wish to select the sub-expression $x^2 - 16$ from $x^2 - x - 16$, and this is often difficult without reordering some of the terms. The unary minus also causes difficulties here, and this is discussed later.

Another form of interaction combines the use of the keyboard with the immediate display in two-dimensional traditional notation. As the user types the keys the expression gradually builds up on the screen. At some intermediate stages the expression is syntactically invalid, and such a mechanism has to pause until a valid fragment can be interpreted. Because of the dynamic nature of this interaction it is difficult to capture and illustrate on a static printed page, so we have not provided an example here. However, the NUMBAS system, Foster *et al.* (2012), provides such an interface within a large range of current web browsers. Any system which replicates or tries to recognize mathematical meaning from traditional notation will have to address these problems. Pen-based entry on tablet machines, e.g. Fujimoto and Suzuki (2003), and handwriting recognition of mathematics face identical challenges.

Many of the design issues are discussed in more detail in Nicaud *et al.* (2004) and Prank (2011), and the majority of editors currently available seem to have addressed them. A review of over twenty contemporary editors was undertaken by Nicaud and Bouhineau (2008) to try to establish whether there is a consensus on how expressions are represented and how such editing actions take place. They concluded that '*existing software has not reached a consensus for presentation, exchange and editing of mathematical expressions. The result is a confusing landscape for users and students trying to learn mathematics with computers.*' Indeed, good design might be characterized by the extent to which users are not aware of the existence of such issues.

5.9 Dynamic interactions

Computers provide opportunities for other interactions which do not result in an algebraic expression. The computer environment is dynamic, and it is perfectly possible to ask a student questions, the responses to which are interactions with dynamic pictures. We provide just two examples, illustrated by the Java applet GeoGebra.

GeoGebra is dynamic mathematics software rather than a computer aided assessment system. It was instigated and developed by Markus Hohenwarter from 2001, with subsequent work by many others including Yves Kreis (Luxembourg) since 2005, and Michael Borcherds (Birmingham, UK) since 2007. It has received several international awards, including the European and German educational software awards. It contains an interactive geometry system in which objects such as lines and conic sections can be constructed and 'dragged' dynamically. It also contains algebraic elements so that equations and coordinates can be entered directly. The combination of algebraic and geometric features is unusual. Essentially, GeoGebra is an exploratory environment for performing mathematical experiments and it is currently difficult to encode a *complete mathematical argument* which includes the logic, rhetoric, and algebraic manipulative steps.

The manipulation of GeoGebra diagrams provides an entirely different mathematical interface from entering an algebraic expression. The question shown in Figure 5.2 is adapted from Sangwin *et al.* (2009). In this question the interaction is quite different from that of a traditional paper-and-pencil geometry problem. As the point P is moved the values of the angles update dynamically, so the student is interacting with a dynamic geometry diagram. The student still needs to work out which lines are parallel and on this basis move P to an appropriate point, so that angle a is exactly $65°$. Furthermore, the immediate feedback provided by the task shown in Figure 5.2 provides a kind of dynamic updated display simply not available on a static piece of paper.

GeoGebra diagrams have also been included in the WeBWork system described in Section 8.9, and a further example is given in Section 8.3 in which a student must solve simultaneous equations by graphical methods. Another important topic in elementary mathematics is *curve sketching*. An example of an interface, also from WeBWork, which

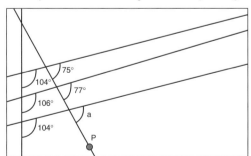

Figure 5.2: Dynamic interactions.

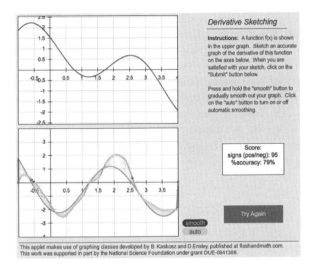

Figure 5.3: Sketching the derivative in WeBWork.

accepts a student's sketch as an answer is shown in Figure 5.3. In this applet, students need to use the mouse to sketch the derivative of a graph shown in the question. The system assesses whether the sketch is sufficiently close to the correct derivative and provides feedback.

None of these example use particularly advanced mathematics. Indeed, the use of graphical methods to solve simultaneous equations deliberately avoids any algebra. However, they do illustrate some options currently available for interacting with mathematical objects, via computer, which do not require the entry of an algebraic expression.

5.10 Conclusion

Mathematical notation has an interesting history and exerts a profound influence over the way we think about the subject. Given the increasing importance of computers, Sangwin and Ramsden (2007) questioned whether the use of a linear syntax might affect written mathematics at an earlier stage. However, as with Brown (1974), history has consistently demonstrated that changes to mathematical notation are exceedingly difficult to achieve. The designer and users of a CAA system are almost certainly stuck, for better or worse, with traditional notation. Have we really progressed since Babbage complained of '*a profusion of notations (when we regard the whole science) which threaten, if not duly corrected, to multiply our difficulties instead of promoting our progress*'? (Babbage, 1827, p. 326)

What is perhaps most shocking is that we still do not have an agreed method for entering elementary mathematical expressions into a machine using a keyboard in a standard unambiguous way. Similarly, there is still no standard way to display mathematics on a computer; Hayes (2009). There is still confusion between surface features of notation and their meaning. For example, is juxtaposition a function application or multiplication? There is also usually no uniform way to distinguish between noun and verb forms of operators.

In this chapter we have concentrated on notation, and how to enter a mathematical expression into a machine. We have chosen to spend less time on the *meaning* of the expressions themselves. Of course, it is artificial to separate the notation from its meaning, but from the point of view of CAA this is not so unnatural: the student enters their expression and the system establishes the meaning. In the next chapter we shall consider the properties of mathematical expressions which a teacher might seek to establish automatically.

6

Computer algebra systems for CAA

Although computer algebra dates back to 1953, the first real discussion of the meaning of equality in computer algebra seems to have been published in 1969, Brown (1969). It is still often necessary to apologize for the apparent difficulty of equality: Steel (2002) says 'Amazingly, testing whether an element is the zero element of the field turns out to be the most difficult operation in the whole scheme!' (Davenport, 2002)

In Chapter 5 we discussed the way that mathematics is written down; that is, the *syntax* of mathematics. This discussion revealed some of the long and rather interesting history, and made explicit some contemporary ambiguities of interpretation. Furthermore, we considered how human–computer interactions facilitate or hinder a student in getting their mathematical expression into a machine. We now turn our attention to the problem of establishing the meaning of an expression, i.e. the *semantics*. Given Chapter 5, it is perhaps not surprising that this is more complex than might first be appreciated.

We readily admit that we confine our attention to simple mathematical expressions. These include rational functions, polynomials, sets, lists, matrices, and systems of equations. Quite which subset a student has encountered will depend on how much mathematics they have learned. Initially, they will be probably be:

1 polynomials,
2 the reciprocal and rational expressions,
3 trigonometric functions, e.g. $\sin(x)$, $\cos(x)$ and $\tan(x)$ in *degrees*,
4 exponentiation (without e), e.g. 2^x and roots,
5 and eventually e^x and logarithms.

Systems of equations are confined to linear, quadratic, and cubic equations in a modest number of variables. The traditional curriculum ordering of these topics is set out by Barnard (1999).

Any computer aided assessment which seeks to establish mathematical properties of expressions of any kind will certainly need to include these. All assessment, including steps in working, proofs, problem-solving, and even project work, relies at crucial moments on the ability to establish that the students' mathematics is 'correct' in some sense. Very often this means establishing some well-defined properties of an expression.

A computer algebra system is software for symbolically manipulating data structures representing mathematical objects. In this chapter we consider in some detail the use of CAS for CAA, and as a result this chapter is more technical. We provide a comparison of different mainstream CAS from the point of view of CAA, and contrast the use of CAS as a sophisticated calculator with the needs of an assessment system.

Since 1995, a number of computer aided assessment systems have used an existing mainstream computer algebra system to provide tools with which to assess mathematical expressions as answers. Some of these are described in Chapter 8. Perhaps the first system to make a mainstream CAS a central feature were Mathwise and AiM, which used Maple. Other systems have access to a different CAS, such as CABLE, Naismith and Sangwin (2004) which uses Axiom, and STACK which uses Maxima; see Chapter 7. From private correspondence, the author is also aware of systems which use Derive and Mathematica in a similar way.

Of course, it is not necessary to use a mainstream CAS to process students' responses in a CAA system, and there are many CAA systems in which a student is required to enter a mathematical answer; see Chapter 8. For designers who prefer particular technologies, a CAS library might not be readily available. As Foster et al. (2012) recently said while developing the NUMBAS system, *'it was necessary to create, from scratch, a computer algebra system entirely in Javascript'*. Our discussion is just as relevant to systems such as these. Indeed, we argue that by developing libraries of mathematical functions they are using 'computer algebra' in its broadest sense.

In all these systems the primary purpose of the CAS is to provide the teacher with a ready-made library of useful mathematical functions. This allows the generation and manipulation of expressions to create structured random questions and corresponding worked solutions. These functions also allow a student's answer to be manipulated and tested objectively against a variety of mathematical criteria. Only once this has been done can various *outcomes* be assigned. While a mainstream CAS provides access to a large and well-supported library of mathematical functions for performing computations, CAA as an application requires quite different functions, usually absent in a mainstream CAS.

When considering a given answer, 'correctness' is usually a combination of a number of individual properties. For example, we may want algebraic equivalence with the teacher's answer *and* for the student's answer to be in factored form. In reality the teacher makes many fine judgements rapidly. Our concern in this chapter is in articulating these properties and discussing the extent to which we may establish them automatically, and how to do so. Only then can we think about how the outcomes should be combined to result in any feedback or numerical marks. When trying to help identify what the student might have done wrong, we could establish equivalence with an incorrect answer arising from a common mistake. This is why we talk of *properties*, such as equivalence with a given expression, not about correctness in an absolute sense.

To keep this chapter to a reasonable length, there are some topics we choose not to detail here, in particular (i) logical expressions in Boolean variables, such as `(A and B) or C`;

(ii) interval arithmetic; and (iii) solutions to systems of inequalities. These are also quite interesting topics mathematically, pedagogically, and from a technical computer algebra point of view.

6.1 The prototype test: equivalence

Our prototype test seeks to establish that the student's answer, assigned internally to a CAS variable SA, is the 'same' as the teacher's expression TA. One approach to this is to use the CAS command `simplify` and evaluate the following pseudo-code:

```
if simplify(SA-TA) = 0 then true else false.
```

When such a test returns `true` we have established that the two expressions are *equivalent*: our prototype property. But what if the test returns `false`? Are we really sure they are not equivalent? The robustness of our test relies on the strength of the particular `simplify` command.

Assuming that both SA and TA contain only a single variable, x, another approach would be to choose random numbers $x_1, \cdots, x_n$ (preferably incommensurate) and evaluate each expression at $x = x_k$ as a floating point approximation. Numerical analysis assures us that a reasonable match between the values for each expression gives a reasonable probability of the two expressions being identical: students' expressions are, after all, usually relatively simple. If we take a dozen or so incommensurate values for x, and a small margin of error, the probability of a false result is very low indeed. In fact, given the difficulties CAS have in reducing trigonometrical expressions to a canonical form, this may be a much more reliable implementation than that of a CAS. This test can be extended to multiple variables. However, if (by chance) all $x_k < 0$ then this test has problems with distinguishing $2x$ from $\sqrt{x^2} - x = |x| - x$. If we choose tidy values, e.g. $x = -2, -1, 0, 1, 2$, then $x^3(5 - x^2)$ looks like $4x$. Having said this, in some situations sampling in this way has positive advantages. For example, if students are asked to approximate a function or estimate a trend-line then tolerances allow for 'fuzzy' testing.

Both tests are actually used, e.g. the first by AiM and STACK, and the second by WeBWork, DIAGNOSYS, CALM, and many others. Do they establish the same thing? Furthermore, often when assessing elementary mathematics, the *form* of the answer is as important as its value. Hence, we seek to establish the written form of an expression just as much as its equivalence to something else. We note that string matching or regular expressions are, except in trivial cases, wholly inadequate, either to establish equivalence or form.

In many cases a student's expression will be manipulated by the CAS before a test is applied. For example, assume a student has been asked to 'give an example of an even function of x'. Correctness is confirmed by substituting $-x$ for x and establishing algebraic equivalence with the original expression. In pseudo-code,

```
if simplify(subst(x=-x, SA)-SA) = 0 then true else false.
```

Here the student's answer is being manipulated mathematically prior to the test being applied and there is no 'correct answer' to which the student's expression is compared. *A mathematical property of the expression itself is established.*

6.2 A comparison of mainstream CAS

To motivate much of what follows, and to convince the reader that there really is an issue to discuss, we shall undertake a very simple comparison of mainstream CAS. Inevitably, a CAA system which incorporates a mainstream CAS accepts the design decisions of the CAS that they adopt. However, the application of CAS to support automatic assessment is quite different from the role of super-calculator to which a CAS is traditionally put. The differences between CAS have been discussed elsewhere; for example, Grabmeier *et al.* (2003) or Wester (1999). However these comparisons are made from the point of view of the research mathematician or computer scientist.

In this section we review four computer algebra systems, Axiom, Maple 9.5, Maxima 5.15.0, and Derive 5.0. The input column of Table 6.1 shows a number of elementary expressions, all of which might occur in learning and teaching. For each CAS we have shown the *default* simplification of these expressions. That is to say, the basic input–output behaviour of the CAS. Notice that there is quite a striking variety of different 'default simplifications', and other design decisions evident in the Table.

We start with numbers. When arithmetic involves both floating point representations and rational numbers a choice has to be made: should the result be a floating point number or a rational? For example, when $\frac{1}{4}$ is subtracted from 0.5, the CAS must either change the rational number to a floating point representation or vice versa. Derive changes floats to rational numbers before completing the arithmetic, while the others choose floating point numbers.

When collecting terms, as in Example 8, Maple, Axiom, and Derive all require the coefficients in a polynomial to have the same data-type. As with numbers, Maple and Axiom default to floats, but Derive uses rational numbers. Notice that Maxima tolerates mixed types within a single polynomial. In Example 8 the constant term is not involved in arithmetic, but Axiom and Maple coerce this into a float. From a strict computer science point of view, it makes sense to have a polynomial over a particular field or algebraic ring. For CAA we need to deal with what the student enters, and since students are apt to enter expressions containing a mix of numerical representations it is necessary to deal with them. Furthermore, while the default in Maxima is to coerce to floats, it is possible to choose coercion to rational coefficients instead.

With surd/radical terms there are two questions to answer: (i) how should the output be displayed, and (ii) what assumptions are made to simplify expressions? In terms of input–output, there are often two methods for entering a square root into the CAS, either using a command such as `sqrt` or using indices such as `2^(1/2)`. In what senses are these the same? It may be important both whether a student enters their expression in a particular way, and whether the system displays an expression using a particular choice of notation, e.g. a traditional $\sqrt{}$ symbol or fractional powers. You might expect a CAS to respect the way

6.2 A COMPARISON OF MAINSTREAM CAS

Table 6.1: A comparison of the elementary behaviour of various CAS.

Ex. No.	Input	Maple	Maxima	Axiom	Derive
	(numbers)				
1	0.5-1/4	0.25	0.25	0.25	$\frac{1}{4}$
2	4^(1/2)	$\sqrt{4}$	2	2	2
3	4^(-1/2)	$\frac{1}{4}\sqrt{4}$	$\frac{1}{2}$	$\frac{1}{2}$	$\frac{1}{2}$
4	(-4)^(1/2)	$\sqrt{-4}$	$2i$	$2\sqrt{-1}$	$2i$
5	sqrt(-4)	$2i$	$2i$	$2\sqrt{-1}$	$2i$
	(indices)				
6	a^n*b*a^m	$a^n b a^m$	$a^{n+m} b$	$b a^m a^n$	$b a^{m+n}$
7	(a^2)^(1/2)	$\sqrt{a^2}$	$|a|$	$\sqrt{a^2}$	$|a|$
	(collecting terms)				
8	x/3+1.5*x+1/3	$1.833x + .333\cdots$	$1.833x + \frac{1}{3}$	$1.833x + 0.333\cdots$	$\frac{11x}{6} + \frac{1}{3}$
9	3*x/4+x/12	$\frac{5}{6}x$	$\frac{5x}{6}$	$\frac{5}{6}x$	$\frac{5x}{6}$
	(brackets)				
10	-1*(x+3)	$-x-3$	$-x-3$	$-x-3$	$-x-3$
11	2*(x+3)	$2x+6$	$2(x+3)$	$2x+6$	$2(x+3)$
12	(2*x-1)/5+(x+3)/2	$\frac{9}{10}x + \frac{13}{10}$	$\frac{2x-1}{5} + \frac{x+3}{2}$	$\frac{9}{10}x + \frac{13}{10}$	$\frac{9x+13}{10}$
	(other)				
13	log(x^2)	$\ln(x^2)$	$2\log(x)$	$\log(x^2)$	$2\ln(|x|)$
14	log(x^y)	$\ln(x^y)$	$y\log(x)$	$\log(x^y)$	$\ln(x^y)$
15	log(exp(x))	$\ln(e^x)$	x	x	x
16	exp(log(x))	x	x	x	x
17	cos(-x)	$\cos(x)$	$\cos(x)$	$\cos(x)$	$\cos(x)$

an expression is entered when it is displayed, but Maxima, for example, converts the input 3^(1/2) to display $\sqrt{3}$.

Example 7 in Table 6.1 illustrates the extent to which the CAS makes assumptions about arbitrary symbols. If we assume that x is real and non-negative, $\sqrt{x^2} = x$; if we assume only that x is real then $\sqrt{x^2} = |x|$. When we make no assumptions about x, $\sqrt{x^2}$ should remain unsimplified, as it does in Maple. Most CAS allow assumptions to be specified to help the simplifier, and Maxima's default behaviour $\sqrt{x^2} = |x|$ can be overridden by a command such as assume(x>0), in which case $\sqrt{x^2}$ is simplified to x.

These assumptions also carry over to the simplification of logarithmic expressions, as shown in Examples 13–16. Different simplification assumptions are necessary to force, for example, $\frac{(n-1)!}{n!}$ to evaluate to $\frac{1}{n}$, or $\cos(n\pi)$ to evaluate to $(-1)^n$. Such assumptions need to be made explicit to the CAS, and may well depend on the question being asked. Certainly for the application of CAA the teacher should be able to specify how the system behaves in fine detail such as this.

When we consider the assumptions made which are relevant during the early stages of learning, mainstream CAS differ significantly from each other. Angular measure is assumed to be in radians, not degrees. The input `log(x)` is interpreted as the *natural logarithm*, although as Example 13 illustrates not all systems display the result this way. However, the notation `log(x)` is often used to denote logarithms to base 10, particularly on calculators. Various simplifications are performed by default on such functions, illustrated in Example 17 where the evenness of cosine is used to simplify $\cos(-x)$ to $\cos(x)$ in all systems.

We have concentrated only on the simplest mathematics, and many more examples are given in Stoutemyer (1991). These issues might at first appear utterly trivial. The expert does not worry about such distinctions: a hallmark of their expertise is that they work effectively regardless of such 'technicalities'. But, during elementary mathematical instruction this is *the point of the work*. In CAA we need to be able to make the kinds of distinction which we have described in this section, and on the basis of these assign outcomes such as feedback.

6.3 The representation of expressions by CAS

While rather technical, we shall consider briefly how a CAS represents an expression internally, beginning with the polynomial $x^3 - 7x + 6$. One way to represent such a polynomial would be as a list of coefficients, e.g.

$$x : [6, -7, 0, 1].$$

Such a representation is known as a *dense form*. An example of a system which takes this approach is the symbolic toolbox of Matlab. Another option, taken by for example Maple, is to represent the data in *sparse form*. In this form we have

| SUM | EXP1 | COEFF1 | EXP2 | COEFF2 | ... |

Each EXPn is itself a Maple expression, such as a power of a variable. Here, only those terms which appear in the sum are represented internally. In both representations the system usually forces all coefficients to be of a particular type, e.g. rational, to give a polynomial in a specific ring. This explains why Maple transforms all the coefficients in Example 8 in Table 6.1. The precise details are given by, for example, Heck (2003).

In both cases we have, from the point of view of CAA, a potential problem. How would we represent an expression such as $x^3 + x + 7 + x$? It could well be, *at the parsing stage* when the student's expression is transformed into an internal representation, that information is lost and we end up with a data structure such as

$$x : [7, 2, 0, 1].$$

Where this is the case, it would be impossible for our CAA system to distinguish between $x^3 + 2x + 7$ and $x^3 + x + 7 + x$. Information is potentially lost in gathering or reordering terms, which might be relevant to the objectives of the question being asked. For elementary CAA these fine-grained distinctions are sometime precisely those required, so the way the CAS represents objects is crucial for us.

In Maxima, any expression, including this polynomial, is represented as a LISP tree. LISP, 'LISt Processing language', relies on data structures which are linked lists. Indeed, Lisp source code is made up of such lists, and this provides a blurring between data and code. Source code can be manipulated as data. This is natural to a mathematician, as a function f can also be considered as both an object and a process. A function is written as a list, with the function's name first and the arguments following. For example, a function f with three arguments might be represented using (f x y z).

Nested lists-of-lists can be used to represent trees. Such trees are very close to traditional mathematics, and are a rather natural way to represent all kinds of expressions. For example, below we show the tree form of the expression $(\ln(x) + 1)^3$.

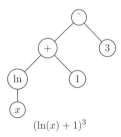

$(\ln(x) + 1)^3$

Even sets can be represented in this way by using a set constructor function, e.g. the set $\{a, b\}$ might be represented as ((SET) a b). Such a 'function' is inert, i.e. it does nothing to its arguments; it just represents a set. From a user's perspective, input is parsed and output displayed as $\{a, b\}$, but internally it still fits within this tree structure. Actually, a set constructor function in a CAS often does process its arguments and it is not entirely inert: duplicates are removed. Furthermore, we shall argue in due course for the importance of inert 'noun' forms and active 'verb' functions for *all* mathematical operations.

When entering an expression into Maxima, the first option is to disable the simplification mechanism completely. Then, typing in the syntax x^3-7*x+6 results in the following internal data structure:

((MPLUS) ((MEXPT) $X 3) ((MMINUS) ((MTIMES) 7 $X)) 6)

The first operation, (MPLUS), represents addition. Since addition is an associative operator, internally it takes an arbitrary length list of arguments. Indeed, we could type this expression into Maxima as `"+"(x^3,-7*x,6)` to reinforce, in our mind at least, the functional nature of addition. The first summand is ((MEXPT) $X 3), itself an expression representing x^3. Even though addition is commutative the system has not changed the order, and minus is treated as a *unary prefix operation*. Notice the subtle differences between $-(7x)$ as we have it here, $(-(7))x$ where minus is a unary operator applied to the positive integer 7 and $-7x$ where 'minus seven' is a legitimate mathematical entity in its own right.

If we keep basic simplification on, the default Maxima setting, we end up with a rather different data structure for our polynomial.

```
((MPLUS SIMP) 6 ((MTIMES SIMP) -7 $X) ((MEXPT SIMP) $X 3))
```

The order of terms has been changed internally, and the integer -7 now exists in its own right. Intriguingly, in both cases the *displayed output* from the CAS is indistinguishable.

Maxima is particularly convenient from our point of view, since it is possible to turn off all simplification using the command `simp:false`. Few other CAS have this ability, which we shall see is vital when making fine distinctions between trees and interpreting their meaning.

Being able to switch simplification on and off is invaluable, but it is a rather blunt tool. Much finer degrees of control over which simplifications are performed are necessary. The work of Gray and Tall (1994) develops the notion of a *procept* to capture the duality between process and concept in mathematics. For example, basic arithmetic operations make use of the same symbolism to represent the product of the process: one half as 1/2, and as the process itself: divide one into two equal parts. They comment on the ambiguities in using the same symbol for both, as follows.

> By using the notation ambiguously to represent either process or product, whichever is convenient at the time, the mathematician manages to encompass both—neatly sidestepping a possible object/process dichotomy. (Gray and Tall, 1994)

The designer of an automated system does not have the luxury of being able to '*use the notation ambiguously*'. Furthermore, to the designer of a mainstream CAS an elementary arithmetic operation, such as +, is always interpreted as *something to do*, i.e. the process. Only when the process cannot be completed is the symbol used by the CAS, for example $1 + x$, to represent the object. These simplifications combine the *algebraic* properties, such as commutativity, with the *functional* properties of the arithmetic operations.

Maxima can suppress all simplification; it is then possible to selectively activate individual operations to perform only additions, or only divisions etc. We follow Maxima's convention of referring to *noun* and *verb* forms of mathematical functions. The noun form corresponds to the concept, and is representative. The verb form corresponds to the process, and this actually performs the work of the function itself. Operators have abstract algebraic properties, such as commutativity and associativity, which are different from the function definition itself. For this application the CAS should enable every function to have both noun and verb forms through one mechanism. Following the conventions of traditional notation, when displayed to students these forms should be indistinguishable.

Parsing exactly the expression entered by the student, and being able to manipulate this with the CAS directly, is vital for CAA. We will need, selectively, to perform the following kinds of operation; Heeren and Jeuring (2009):

- Switch on associativity to remove 'unnecessary' parentheses.
- Switch on commutativity and to tell the CAS to reorder terms.
- Gather like terms in a sum, effectively performing selective additions.
- Perform selective multiplications, including the use of exponential notation.

How expressions and sub-expressions are represented affects how parts can be selected and therefore how substitutions are performed. The most basic objects are *atoms*, e.g. integers, floating point numbers, and unassigned variables. Everything else is a tree. With such a tree structure the most significant *operation* (op(ex)) and the list of arguments (args(ex)) form the basis for such substitutions.

Let us consider the *expanded form* of a polynomial. With a single variable we traditionally just order terms in descending powers of the variable. However, in the multivariate case it is not clear whether x^2y or y^2x should appear first in the sum. For cubic polynomials in two variables x and y we have terms of the form $x^n y^m$, where n and m take values 0, 1, 2, 3, and $n + m \leq 3$. Specifically, we have the following types of term, each with an appropriate coefficient, if we expand out our cubic.

	1	x	x^2	x^3
1	1	x	x^2	x^3
y	y	xy	$x^2 y$	
y^2	y^2	xy^2		
y^3	y^3			

To decide how to write the cubic we start by fixing a *term order*, a total order $<$ on the set of all possible power products, i.e. products of variables raised to powers. Notice that even when writing the terms in this table we have ordered the variables within the term, writing x^2y and not yx^2. The order is a mechanism for writing such terms as a one-dimensional string, which can be thought of as a way of traversing this table. One example is to write a polynomial as

$$(.x^3 + .x^2y + .xy^2 + .y^3) + (.x^2 + .xy + .y^2) + (.x + .y) + ..$$

Here the . represents the coefficient. The parentheses are unnecessary, but group terms $x^n y^m$ for which $n + m$ are equal. Within these groups we have ordered by highest coefficient of x. There are many different orders possible with multi-variable polynomials, and for some techniques, e.g. calculating a Gröbner basis, this is significant.

These orders are used whenever we write a sum of terms, for example when we write the polynomial in expanded form, e.g. $x^2y^2 - y^2 - x^2 + 1$. We might want to operate on the whole expression, for example, to factor it completely:

$$x^2y^2 - y^2 - x^2 + 1 = (x-1)(x+1)(y-1)(y+1).$$

Equally, we might think of this as a polynomial in x with coefficients in y,

$$x^2(y^2 - 1) - y^2 + 1.$$

Lastly, we might need to see the coefficients of this polynomial in factored form.

$$x^2(y-1)(y+1) - (y-1)(y+1).$$

In mathematics such manipulations are rather common, and yet in some CAS they are very difficult to achieve. One exception to this is Maxima's `format()` function contributed by Miller (1995). This function allows the user to recursively arrange the expression according to a chain of templates. The templates are described in a semantic, indeed almost algebraic, way. If p is our polynomial, then the three forms of p are obtained by the following code:

```
1  format(p,%factor);
2  format(p,%poly(x));
3  format(p,%poly(x),%factor);
```

By describing such manipulations as *formatting*, we keep firmly in mind the idea that each is a form of the same mathematical object.

While the precise internal details should probably be unseen by the question author, it is very helpful if they appreciate these issues, especially if they intend to write sophisticated questions making fine distinctions between forms of expressions.

6.4 Existence of mathematical objects

While we might argue, at a philosophical level, whether mathematical objects *exist*, we can hopefully agree that humans have mathematical objects in mind, whereas computers represent them internally with a formal data structure. One common interface from mind to computer is *syntax*, and from computer to human is *display*. Some examples are given in Figure 6.1.

We shall assume, for the purposes of the rest of this chapter, that an appropriate interface exists, and that internally the machine has *expression trees* which represent the mathematical expressions of interest. We shall assume that the parsing respects the intentions of users entering their expressions (for example, student and teacher), but acknowledge that getting

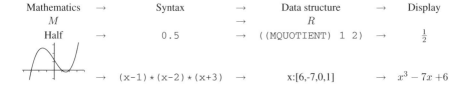

Figure 6.1: Mathematical objects and their representation.

to this stage is far from simple. Note that the parsing process itself requires syntactical validity from some user interface.

Our focus is in establishing the mathematical properties of the objects which these expression trees represent. In particular, we shall assume that the process of parsing the expression performs no *simplification* whatsoever. For this application the teacher really needs *exactly the expression provided by the student in a machine-manipulable form*; hence the focus on expression trees.

We need a little care here when thinking about mathematical objects and their representations. For example, do we really believe that $\frac{x^2-1}{x-1} = x + 1$? First consider functions, $\mathbb{R} \to \mathbb{R}$ defined by the formulae $f_1(x) = \frac{x^2-1}{x-1}$ and $f_2(x) = x + 1$. While f_1 is undefined at $x = 1$, f_2 is perfectly well defined at this point: they must be different functions. But, as elements of the ring $\mathbb{Q}(x)$, of rational expressions, $\frac{x^2-1}{x-1}$ is considered equal to $x + 1$. Furthermore, since $\lim_{x \to 1} f_1(x) = \lim_{x \to 1} f_2(x) = 2$, the *limiting behaviours* of each function correspond. Blurring the difference between the expression as the definition of a function (analysis) and the expression as an element in a ring (algebra) is common in mathematics. It is another example of the deliberate ambiguity which gave rise to the verb/noun distinction, and at times it can be helpful. So should we just ignore this technicality? Elementary mathematics instruction sometimes seems to do so, just as most texts on elementary mathematics assert $\sqrt{x^2} = x$ and fail to make the restriction $x \geq 0$ explicit.

More formally, following Davenport (2002), let us imagine some abstract mathematics objects M. A surjective function $f : R \to M$ is called a *representation* of M by R. That is to say, for each element $m \in M$ there exists at least one element of R, and for every element of R there is one and only one element of M. We shall, for the purpose of illustration, assume R is the set of Maxima expression trees. Students communicate their mathematics in terms which end up as elements $r \in R$. Let $R' \subset R$. If $f : R' \to M$ is a bijection then the representations of R' are *canonical*.

There are essentially two distinct jobs for the CAS to perform.

1 *Establish mathematical properties of $f(r)$*
   Does the student's answer represent a mathematical object which has the correct properties?

2 *Given $R' \subset R$, establish membership $r \in R'$*
   In many cases this amounts to establishing if r has the correct syntactic form, whether canonical or not.

The sets

$$R_m : \{r \in R : f(r) = m\}$$

are equivalence classes. A function which returns a unique element of each R_m is called a *canonical form*. In mathematics it is relatively common for at least part of a problem to require a student to transform a mathematical representation $r_0 \in R$ into $r_n \in R'$ via a sequence of 'steps'. For each step r_i, $i = 1, \cdots, n$ we have $f(r_0) = f(r_i)$. i.e., we transform the representation of a particular mathematical object. This is called *reasoning by equivalence*, and is precisely the task for which Aplusix, discussed in Section 8.11, was designed. For example, in the question 'solve the equation $4x + 7 = 15$', the student might create the sequence of steps $4x + 7 = 15$, $4x = 8$, $x = 2$. Each of these represents the same equation, the last of which is in canonical form. However, it is probably acceptable, in this case, for the student to enter $x = 2$ or $2 = x$, so that the teacher does not require the canonical form, but equivalence to the canonical form respecting the commutativity of the equality sign.

The quotation at the start of this chapter suggests that testing whether something is the zero element is difficult. But why might this be the case? CAS deal with infinite precision arithmetic, and with algebraic numbers. Simple surd expressions have many forms; e.g. Beevers *et al.* (1991, p. 78) gives the following forms of $3 - 2\sqrt{2}$:

$$(\sqrt{2} - 1)^2, \quad \frac{\sqrt{2} - 1}{\sqrt{2} + 1}, \quad \frac{1}{(\sqrt{2} + 1)^2}, \quad \frac{1}{3 + 2\sqrt{2}}.$$

More challenging examples occur when the surds are nested as in the first expression below. Let us, for the moment, compare

$$\sqrt{11 + 6\sqrt{2}} \text{ with } 3 + \sqrt{2}.$$

In fact, these numbers are the same, so their difference is zero. One technique for *de-nesting surds* is given by (Euler, 2006, §671–81) which amounts to defining $c^2 = a^2 - b$ and using the identity

$$\sqrt{a + \sqrt{b}} = \sqrt{\frac{a + \sqrt{a^2 - b}}{2}} + \sqrt{\frac{a - \sqrt{a^2 - b}}{2}} = \sqrt{\frac{a + c}{2}} + \sqrt{\frac{a - c}{2}}.$$

When c is a perfect square this results in a genuine simplification, as in our example. There are similar identities for forms such as $\sqrt{\sqrt{a} + \sqrt{b}}$ and for other fractional powers. The challenge, therefore, is to develop (i) a canonical form for algebraic numbers and (ii) effective algorithms for transforming a particular number into this form. Further comments on this problem are given by Wester (1999, Chapter 4) and Davenport *et al.* (1993), and the general problem is solved in Landau (1992).

It is interesting to note that the need to denest surds is rather common: solving a quadratic equation with surd coefficients using the formula, or applying Cardano's formula for the solutions to the cubic equation with integer roots, both result in nested radicals. For example, in *Ars Magna* (Cardano (1993)) Gerolamo Cardano (1501–1576) solves the equation $x^3 + 6x = 20$, written in modern notation. Since $x^3 + 6x$ is a bijection of the reals the equation has one real solution, and it is simple to confirm this equals 2. Cardano's formula results in

$$\sqrt[3]{\sqrt{108} + 10} - \sqrt[3]{\sqrt{108} - 10}$$

which is clearly a positive real number, and this must therefore equal 2. Such expressions also arise when using the CAS to solve equations internally, e.g. when calculating the eigenvalues of a matrix or automatically generating steps in a worked solution, and hence this issue cannot be avoided, even in elementary mathematics. However, currently United Kingdom mathematics curricula do not include the technique of de-nesting surds. This was an important topic and algebra texts such as Chrystal (1893) devote considerable space to these techniques. They appeared regularly in examinations, e.g. Austin (1880). Indeed, Euler (2006) introduces addition, multiplication etc. of *compound terms* such as $1 + \sqrt{3}$ before algebra with abstract symbols. Manipulation of such compound expressions precedes solving equations with symbols such as x.

Interestingly, from a theoretical perspective, it is not possible to construct a general algorithm to test the equivalence of arbitrary expressions. Originally, Richardson (1966) showed that there is no algorithm which can establish that a given expression is zero in a finite number of steps. His function class included exponentiation, but this result was refined by Matiyasevich (1993) to encompass the following *elementary function class*. Take constants from the integers extended by π, and the single variable x. Form the closure of expressions constructed using addition, multiplication, division, substitution, sine, and the modulus function. Then, there is no algorithm for deciding if such an expression is zero. Hence, there is no algorithm for deciding if two expressions are equivalent, i.e. their difference is zero. If we take the simpler class of polynomials over the complex rational expressions together with unnested exponential functions, then Caviness (1970) showed we do have a decision procedure. Hence in this situation we can be confident if an expression is equivalent to zero. Moses (1971) comments:

> In fact, the unsolvability problem may lie in Richardson's use of the absolute value function. When one adds the absolute value function to a class of functions which forms a field (e.g. the rational functions), then one introduced zero divisors. For example, $(x + |x|)(x - |x|) = 0$, although neither factor is 0.

While Richardson's result is a theoretical restriction on the effectiveness of our test for algebraic equivalence, in practice, for learning and teaching such tests work very well indeed on the limited range of expressions used. Further discussion and examples are given by Beeson (2003). As Fenichel (1966) comments '*recursive undecidability can be a remote and unthreatening form of hopelessness*'.

6.5 'Simplify' is an ambiguous instruction

Notice that implementing the prototype equivalence test relies on the CAS command `simplify`, but what does it do? When using a computer algebra system, this is a very commonly used command. We have used this extensively in our discussion. However, we now argue that this is ambiguous, and indeed it can be and regularly is used for the opposite mathematical operations.

For example, 1 is simpler than 2^0, but $2^{2^{10}}$ is probably simpler than writing the integer it represents in decimals. Everyone would agree that $x + 2$ is simpler than $\frac{x^2-4}{x-2}$, but we might argue that the first expression below is simpler:

$$\frac{x^{12} - 1}{x - 1} = x^{11} + x^{10} + x^9 + x^8 + x^7 + x^6 + x^5 + x^4 + x^3 + x^2 + x + 1.$$

These examples can easily be adapted to factored or expanded forms. Similar examples can be found for many other pairs of forms, e.g. single fractions and partial fractions, or various trigonometrical forms. It is not difficult to find examples in textbooks. In Tuckey (1904) the word 'simplify' is usually taken to mean (e.g., p. 11, Ex 8) *'simplify by removing brackets and collecting like terms'*. Arguably, 'removing' should be 'expanding' here. But 'simplify' is also later used to implicitly mean *factor and cancel like terms*. For example,

$$\text{(Tuckey, 1904, p. 139, Ex 77). Simplify } \frac{a^4 + a^2b^2 + b^4}{a^3 - b^3}.$$

This is typical[1] of contemporary usage. The word 'simplify' is so vague that its use is generally very unhelpful indeed. It may mean little more than 'do what I have just shown you'. What it certainly does mean is to *transform* an expression into an equivalent, but different form.

Fitch (1973) gave three reasons for simplifying expressions. The last of these was the *identity problem*; that is, to see if an expression is identically zero. We have already commented on this issue, which is central to CAA. The first is what he calls *compactness* of expressions, to make the expression smaller. This idea can be found in older writers; for example, Babbage (1827, p. 339) comments that

> whenever in the course of any reasoning the actual execution of operations would add to the length of the formula, it is preferable to merely indicate them.

By 'indicate' Babbage means the noun form of an operator. Or further back (Euler, 1990, §50),

> [...] the simplicity of the equation expressing the curve, in the sense of the number of terms.

---

1. C. O. Tuckey, author of Tuckey (1904), was a very well respected teacher, president of the Mathematical Association, and author of many books and widely circulated reports (e.g. Tuckey (1934)) into effective teaching. This example is not a personal criticism, but rather an example of typical usage.

Designers of contemporary CAA have also reached this conclusion, e.g. Beevers et al. (1991, p. 113):

> It has long been accepted in science that 'the simplest answer is the right one'. We have translated this premise into 'the shortest answer is the right one'.

Expanding on this idea, Carette (2004) proposed the following informal definition.

Definition 1 *An expression A is simpler than an expression B if*

- *in all contexts where A and B can be used, they mean the same thing, and*
- *the length of the description of A is shorter than the length of the description of B.*

Formalizing this using complexity theory, Carette (2004) applies these ideas to the binary encodings of the integers and operations +, ×, −, and exponentiation. He argued that 2^7 is more complex than 128, but that 2^8 is simpler than 256. However, given two explicit integers n and m 'it is never the case that the algebraic expression $n + m$ is simpler than the integer q equal to $n + m$.' This view of simplification is based on a measure of the representational complexity of the expression. Simplification is then an attempt to represent the expression in a form with minimal value for this measure. We should note that compactness is strongly related to the way in which information is represented: a discussion of single-variable polynomials is given by Davenport and Carette (2009).

Related to storage space, complexity can also be interpreted in terms of the ease with which computations can be performed on an expression. This view was developed by Moses (1971):

> Of course the prevalence in algebraic manipulation systems of simplification transformations which produce smaller expressions is due mostly to the fact that small expressions are generally easier to manipulate than larger ones.

The second reason Fitch (1973) gives for simplifying expressions, also discussed by Fenichel (1966), is *intelligibility*; that is, making it easier for users to understand. It is not immediately clear that compactness and intelligibility are different. As one example, consider the replacing trigonometric functions by complex exponentials

$$\sin(x) = -\frac{i}{2}\left(e^{ix} - e^{-ix}\right), \quad \cos(x) = \frac{1}{2}\left(e^{ix} + e^{-ix}\right).$$

Using these we remove the need for any trigonometric identities. The rules $e^x e^y = e^{x+y}$, $(e^x)^y = e^{xy}$ and $e^0 = 1$ suffice. In this process we also remove the redundancy in using tan, cosec etc. and a plethora of separate rules. An alternative strategy is to introduce a new variable using the *Weierstrass substitution* $t = \tan(x/2)$. Since

$$\sin(x) = \frac{2t}{1+t^2} \quad \text{and} \quad \cos(x) = \frac{1-t^2}{1+t^2}$$

we have reduced these trigonometrical functions to rational functions of t. Now the expression becomes a polynomial identity and we have reduced the problem to that of deciding if two rational polynomials are identical. It is intriguing that we maintain tan, etc., while other trigonometric functions such as the *versed-sine* $\text{versin}(x) = 1 - \cos(x)$ have long fallen out of common usage. Hence, these transformations render expressions much easier for the machine to manipulate, with fewer rules and fewer operations. A user, on the other hand, may expect their answer in terms of these traditional trigonometric forms rather than as complex exponentials. *Compactness* and *intelligibility* are different issues.

A more important argument for intelligible expressions is that they enable humans to recognize patterns, or structure. Rewrite rules, such as $x \times 1 \to x$ and $x \times 0 \to 0$, which are performed too soon might obscure a pattern. They might simply be wrong; for example, if x subsequently evaluates to $\frac{1}{0}$ then we need to use L'Hopital's rule to evaluate $x \times 0$, not the rule $x \times 0 \to 0$.

Given the potential ambiguity in a meaning for 'simplify' we shall try to avoid using this word again from this point onwards, and instead develop a much more sophisticated vocabulary with which to talk about algebraic operations and the senses in which two expressions can be compared. Others, e.g. Kirshner (1989), agree with this need:

> This analysis, we believe, points the way to a new pedagogical approach for elementary algebra, an approach that requires syntactic and transformational processes to be articulated declaratively, enabling more, rather than fewer, students to escape from the notational seductions of non-reflective visual pattern matching. (Kirshner, 1989, p. 248)

However, completely avoiding using 'simplify' has proved impossible and occasionally use has crept back in.

6.6 Equality, equivalence, and sameness

Our prototype test seeks to establish *algebraic equivalence*. This is implemented with either of the approaches of Section 6.1. However, there are a number of quite different types of equivalence which it is very useful to establish when assessing students' work. The following tests have been found, in practice, to be the most useful senses of equality, while being technically feasible to implement. We have an increasing strictness of interpretation of when two objects are considered to be the 'same'.

Same 'type' of object

In formal computer science the type of an object has a very specific sense. In elementary mathematics there are a number of types of mathematical objects:

- elementary expressions, e.g. polynomials, rational expressions, or expressions with $\cos$, e^x etc.

- equations and inequalities,
- sets and lists,
- matrices.

We need a test to establish whether two expressions are of the same type. It would make no sense to compare a set with a list. Attempting to subtract a number from a matrix, in order to 'simplify' the result and compare with zero, should throw a type error: you can only perform subtraction on terms of the same type. While the result of such a test should return false, we would want to make a distinction between types of object so as to provide feedback which helps the student understand that the type of their answer differs from that of the teacher's. It is relatively common for students to provide an expression $x^2 + 2x + 1$ when the teacher expects an *equation* $x^2 + 2x + 1 = 0$.

Such a test works recursively over the entire expression, so a list of equations is different from a list of polynomials. In order to provide feedback, it acts recursively on objects such as sets and lists to identify which members differ in type. Matrices are checked for size, and corresponding matrix elements are examined individually.

Although we have mentioned lists and sets there are arguably three mathematical structures here: lists, bags, and sets. Lists are ordered and may contain repetition, whereas sets are unordered without repetition. Bags are unordered and may contain repetition. However, bags are not commonly used in mathematics, which is a pity, since it would be more useful to talk about the 'bag of solutions' rather than the 'set of solutions' to capture repeated roots elegantly. We also have 'bags of factors'. Notice that in sets, duplicates are removed. It is important to establish which notion of 'sameness' is applied.

The numbers within these mathematical objects may have types of *rational* and *float*. Surds and mathematical constants π, e, γ should also be considered as 'numbers', even if from a formal point of view they are atomic CAS symbols, or operators and arguments, rather than numeric datatypes. Is a complex number $a + ib$ a single entity or a sum? Back to Gray and Tall (1994)'s ambiguity again. When it is necessary to coerce numbers to the same type (e.g. when actually adding 0.33 to 1/2) the mathematical point of view would suggest that floats should be coerced to rational numbers in most situations: after all, finite decimal numbers are rational numbers. While this might look strange to students at first, the author believes that this reinforces the intended meaning of decimal notation.

Substitution equivalence

Let us now consider a slightly stronger form of equivalence. To do this, consider, for example, a situation where a student types in $X^2 + 1$ rather than $x^2 + 1$. In this case we could establish algebraic equivalence by using case insensitivity. However, given two expressions ex1 and ex2, we could also seek a substitution (i.e. relabelling) of the variables of ex2 into those of ex1, which renders ex1 algebraically equivalent to ex2. If ex1= $X^2 + 1$ and ex2= $x^2 + 1$ then for our example, the required substitution is $X = x$. This test is surprisingly useful, especially in establishing whether the student has used the wrong variable name beyond case insensitivity. The teacher might also ask the student to write an equation representing a situation without wanting to give explicit variable names in advance.

In a CAA context there are many situations where a practical teacher would condone such an error, with comment of course. It might be very confusing to students why their answer is marked incorrect. In practice, the test is applied only if the number of variables in each expression is equal and is small (e.g. less than five). Notice that here we have used algebraic equivalence as the notion of sameness. We could seek to establish another kind of equivalence after substitution.

Beyond this are more general pattern matching tests, which are referred to in computer science as *unification*. The unification process finds values of variables which ensure that two object descriptions can be made equal. See Bundy (1983).

Algebraic equivalence

This is the prototype test, as discussed in detail in Section 6.1.

Associativity and commutativity

This test seeks to establish whether two expressions are the same when the basic arithmetic operations of addition and multiplication are assumed to be nouns but are commutative and associative. Hence, $2x + y = y + 2x$ but $x + x + y \neq 2x + y$. The real difficulties here are the inverse operations, and in particular the unary minus. One approach is to treat $-x$ as 'minus one times' x. Here Maxima would represent $-x$ as ((MTIMES SIMP) -1 $X). But in this case we cannot distinguish between $-x$ and $-1 \times x$. Another approach, taken by Maxima with `simp:false`, is to treat the unary minus as a function, literally having $-x$ represented as ((MMINUS) $X). Now we have difficulties with $-ab + c$ and $c - ab$.

$-ab + c$ is represented as ((MPLUS) ((MTIMES) ((MMINUS) $A) $B) $C)

whereas

$c - ab$ is represented as ((MPLUS) $C ((MMINUS) ((MTIMES) $A $B))).

Even when MTIMES and MPLUS are defined to be commutative and associative these expressions cannot be equal.

Instead, we define a new constant UMINUS to represent the unary minus. Instead of representing unary minus as a *function*, we represent it as multiplication by this special constant. In practice we replace ((MMINUS) ex) with ((MTIMES) (UMINUS) ex). By defining multiplication to be associative the trees flatten, × becomes a so-called *n*-ary operator, and in each case the representation becomes

$-ab + c$ is represented as ((MPLUS) ((MTIMES) (UMINUS) $A $B) $C),

$c - ab$ is represented as ((MPLUS) $C ((MTIMES) (UMINUS) $A $B)).

Now, additional commutativity of + and × are sufficient to establish the equivalence of these expressions. For clarity the tree structures representing three of these expressions are shown here.

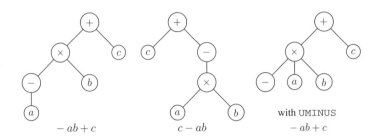

In this approach the unary minus commutes with multiplication in a way natural to establishing the required form of equivalence. Notice that this test does not include laws of indices, so $x \times x \neq x^2$. Since we are dealing only with nouns $- \times -$ does not 'simplify' to 1. E.g. $-x \times -x \neq x \times x \neq x^2$. An extra rewrite rule *could* be added to achieve this, which would change the equivalence classes.

A similar approach is taken with division. Internally in many CAS, division is removed by transforming a/b to $a \times b^{-1}$.

Equivalence up to commutativity and associativity of addition and multiplication of the elementary operations is a particularly useful test for checking that an answer is the 'same' and 'simplified'.

Expression tree equality

This ensures that the two expressions have the same representation in the data structure of the CAS. Literally, we perform a string match on the internal representation of the data. This is the strictest notion of all. For example, the expressions $x + y$ and $y + x$ have different representations as trees, but in few situations would a teacher accept one but not the other. In practice this test is rarely helpful, which is why string matching and regular expressions are so limited.

6.7 Forms of elementary mathematical expression

So far we have concentrated on comparing two expressions. However, it is also important that we can establish that an expression is in a particular *form*. For example, that a quadratic is *factored* or in a *completed square* form. In practice the most useful tests are not equality with a *canonical form*. This is usually too strict a test in learning and teaching. Instead we need to consider and develop something more subtle. We shall provide only a few illustrative examples in this section.

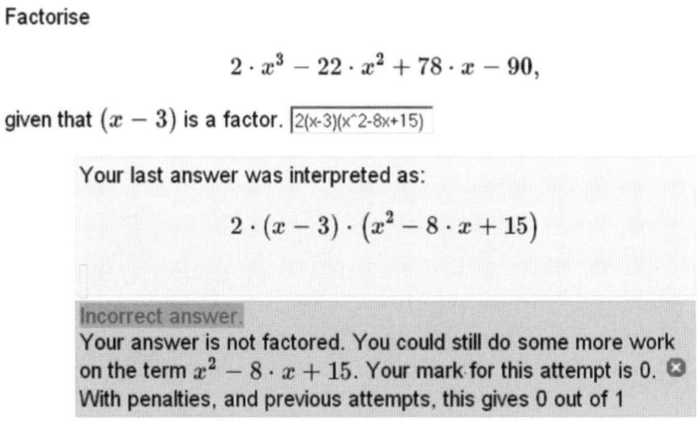

Factorise
$$2 \cdot x^3 - 22 \cdot x^2 + 78 \cdot x - 90,$$
given that $(x-3)$ is a factor. $\boxed{\text{2(x-3)(x\textasciicircum 2-8x+15)}}$

Your last answer was interpreted as:
$$2 \cdot (x-3) \cdot (x^2 - 8 \cdot x + 15)$$

Incorrect answer.
Your answer is not factored. You could still do some more work on the term $x^2 - 8 \cdot x + 15$. Your mark for this attempt is 0. ✗
With penalties, and previous attempts, this gives 0 out of 1

Figure 6.2: Feedback about the form of an answer.

Intriguingly, few mainstream CAS are able to establish forms. While they all have a `factor()` command, few have an `is_factored()` predicate function, returning true or false. In any case, when the test returns false we would also want the test to provide feedback to the student explaining why. Consider the feedback given in Figure 6.2. Here the term in the product which can be factored further is identified for the student. Of course, this feedback may not be suitable for some groups of students. In Section 6.11 we shall return to the *outcomes* it has been found helpful for a test to return.

It is normal to seek to establish that numbers occurring within an expression are acceptable. The most common issues are (i) use of floating point numbers rather than rational numbers, and (ii) rational numbers not written in lowest terms. This is not so easy to establish. As we have already seen in Table 6.1, there is a real difference between $\frac{5x}{6}$ and $\frac{5}{6}x$.

Expanded vs factored

Two important forms in mathematics are expanded and factored. CAS are adept at manipulating expressions to format them in these ways. Note that checking whether an expression is factored is significantly different from comparing an expression `ex` with the result of `factor(ex)`. Consider the following forms of $x^2 - 4x + 4$:

$$(x-2)(x-2), \quad (x-2)^2, \quad (2-x)^2, \quad 4\left(1 - \frac{x}{2}\right)^2.$$

One might argue that each of these is factored.

An expression is said to be factored if it is written as a product of powers of distinct irreducible terms known as factors. Strictly speaking, in establishing that an expression is in factored form, we might not even care whether the terms in the product are 'fully simplified', as long as they are irreducible.

Irreducibility, on the other hand, means that we cannot find further factors, but here we need some care. Bradford et al. (2010) identified the following meanings, illustrated by $x^8 + 16x^4 + 48$.

1. Any non-trivial factorization, e.g. $(x^4 + 4)(x^4 + 12)$.
2. A factorization into irreducible factors over the integers,
   i.e. $(x^2 + 2x + 2)(x^2 - 2x + 2)(x^4 + 12)$.
3. A factorization into terms irreducible over the reals,
   i.e. $(x^2 + 2x + 2)(x^2 - 2x + 2)(x^2 + 2\sqrt[4]{3}x + 2\sqrt[4]{3})(x^2 - 2\sqrt[4]{3}x + 2\sqrt[4]{3})$.
4. A factorization into irreducible polynomials over the Gaussian integers, with i allowed; i.e. $(x + 1 + i)(x + 1 - i)(x - 1 + i)(x - 1 - i)(x^4 + 12)$.
5. A factorization over the complex numbers, where the factor $(x^4 + 12)$ would also be split into the four terms $x \pm \sqrt[4]{3}(1 \pm i)$.

In elementary teaching, meaning 4 is unlikely to occur. Indeed, we might take this example to represent factoring over an arbitrary ring or extension field of $\mathbb{Q}$. We normally seek to establish that the factors are irreducible over the integers (which is equivalent to irreducibility over the rational numbers) or the reals. But, unlike a canonical form, we are not particularly interested in the order of the terms in this product, or the order of summands inside these terms.

There are some delicate cases, such as $(2 - x)(3 - x)$ vs $(x - 2)(x - 3)$ and $(1 - x)^2$ vs $(x - 1)^2$. Nicaud et al. (2004) also make interesting observations on this topic. They define two concepts: *P-factored* and *N-factored*. P 'for polynomial' is applicable to polynomials in one variable, and ensures that the expression is written as a product of '*prime polynomials*' over the reals. N 'for numerical' factored is limited to expanded polynomials of one variable and indicates if numbers have been factored.

There is not a unique way to define N-factored; for example:

- Is $2x^2 + x/2$ N-factored or do we need to factor it into $\frac{1}{2}(4x^2 + x)$ or into $2(x^2 + x/4)$?
- Is $2x + \sqrt{2}$ N-factored or do we need to factor it into $\sqrt{2}(\sqrt{2}x + 1)$?

Numerical factors are clear at a syntactic level when the same numerical expression is a common factor, e.g., in

$$(2 + \sqrt{2})x + (2 + \sqrt{2}).$$

Numerical factors are also clear at a semantic level over the integers, e.g., in $15x^2 + 6x - 9$. We chose to combine these two features to implement the N-factored property in APLUSIX. (Nicaud et al., 2004)

This quotation indicates the real choices which need to be made in design, and when teaching. What does *factored* actually mean? The teacher probably requires more than one predicate to establish different notions of 'factored'.

Establishing that an expression, ex, is expanded is much more straightforward. Essentially, we compare ex with expand(ex) up to commutativity and associativity of the algebraic operations.

Rational expression vs partial fraction

Testing for a rational expression is relatively simple. We need to establish that the denominator and numerator have no common factors, otherwise feedback should be available.

Partial fractions form is more difficult to recognize. Just as with the factor test this is significantly different from checking equivalence with the result of the partfrac function. There are also subtleties here, as illustrated by

$$\frac{1}{n+1} + \frac{1}{1-n} = \frac{1}{n+1} - \frac{1}{n-1}$$

and

$$\frac{1}{4n-2} - \frac{1}{4n+2} = \frac{n}{2n-1} - \frac{n+1}{2n+1}.$$

6.8 Equations, inequalities, and systems of equations

When are two equations the 'same'? Assume we wish to compare two equations, $l_1 = r_1$ and $l_2 = r_2$ say. One rather strict notion would be to compare l_1 with l_2 and r_1 with r_2. A less strict test would be to consider whether $(l_1 - r_1) = \pm(l_2 - r_2)$. Another approach is to check $l_2 - r_2 \neq 0$ and compute $\frac{l_1 - r_1}{l_2 - r_2}$ as a canonical rational expression, i.e. 'simplify' it. If this ratio simplifies to a number then the equations are the same. A last approach is to write n_1 as the numerator of the rational expression $l_1 - r_1$, and n_2 as the numerator of $l_2 - r_2$. If $n_2 \neq 0$ and n_1/n_2 simplifies to a number then the equations are the same.

Systems of equations require an altogether more sophisticated approach. To illustrate this we consider an example adapted from Ex 61.19 of Tuckey (1904).

▼ Example question 13

In a railway journey of 90 kilometres an increase of 5 kilometres per hour in the speed decreases the time taken by 15 minutes. What is the speed?

6.8 EQUATIONS, INEQUALITIES, AND SYSTEMS OF EQUATIONS

This algebra story problem belongs to the distance, rate, and time family: $d = v \times t$ in the classification system of Mayer (1981). It is a further example of a word problem, as discussed in Section 3.5.

In answering such problems, Sangwin (2011b) argued that the first step involves an important early form of *modelling*. In this case there is a mix of units of time between hours and minutes in the statement of the question which calls for careful attention. Taking v and t to be the original speed and time (in hours) we have the system of equations

$$\{90 = vt, 90 = (v + 5)(t - 1/4)\}. \tag{6.1}$$

The next step would be to eliminate t from the second equation, ultimately reducing it to the quadratic $v^2 + 5v - 1800 = 0$ which can readily be solved. It is likely that such a problem will be posed as a multiple-step CAA question in which the system of equations (6.1) forms part of the question. An example, from STACK, is shown in Figure 6.3. When we ask a student to model this situation by writing down a system of equations that describes it, they may not choose the same equations as we do. What, then, does it mean for two systems of equations to be the 'same'? One option is to look for the same *set of equations*, where we reduce sameness to that for equations, but actually this notion is too strict. For example, we could have a set such as

$$\{t = 90/v, 90 = (v + 5)(90/v - 1/4)\}.$$

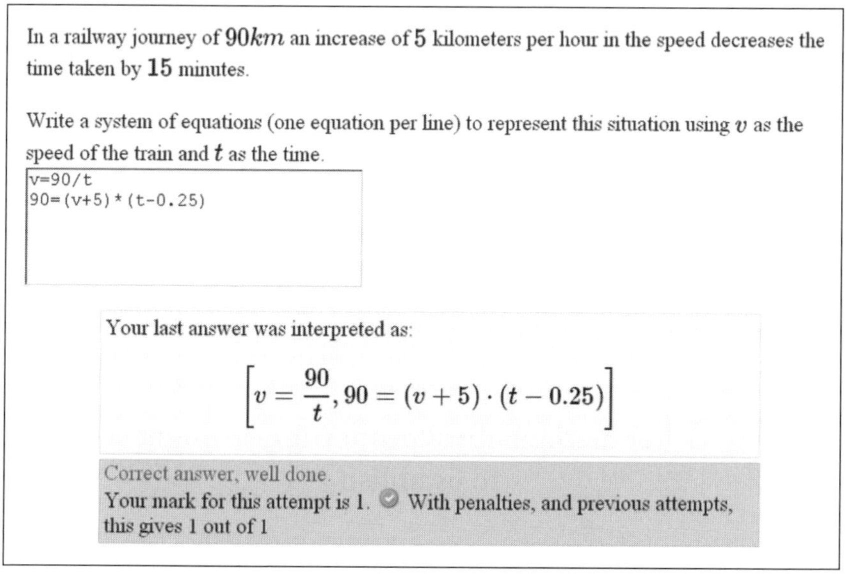

In a railway journey of $90km$ an increase of 5 kilometers per hour in the speed decreases the time taken by 15 minutes.

Write a system of equations (one equation per line) to represent this situation using v as the speed of the train and t as the time.

```
v=90/t
90=(v+5)*(t-0.25)
```

Your last answer was interpreted as:

$$\left[v = \frac{90}{t}, 90 = (v + 5) \cdot (t - 0.25) \right]$$

Correct answer, well done.
Your mark for this attempt is 1. ✓ With penalties, and previous attempts, this gives 1 out of 1

Figure 6.3: An answer which is a system of equations.

This clearly represents the same situation as (6.1), and while the first equations can be recognized as the same, the second equations are clearly not, because a substitution has taken place. Hence, simply looking for a set of equations is far too strict.

Two systems of equations are 'the same' when they describe the same situation; that is to say, they have the same solutions. The set of solutions of such a system

$$\{e_1 = 0, e_2 = 0, \ldots, e_n = 0\}$$

is a set of assignments of the variables, $x_1, \ldots, x_k$, in the equations such that each equation holds. This set of solutions is called the *variety* of the system. When using this notion of sameness, two sets do not even have to have the same number of equations in each.

There are two situations in elementary mathematics where the variety is easy to calculate. If we have a system of polynomial equations in a single variable, then the variety is the set of solutions to the highest common factor of $\{e_1, e_2, \ldots, e_n\}$. This can be calculated effectively using the *Euclidean algorithm*. If we have a system of linear equations in k variables, then *Gaussian elimination* can be used to 'solve' them.

The railway problem falls somewhere between these two extremes. Fortunately, *Buchberger's algorithm* can be used to calculate the reduced polynomial Gröbner basis, Adams and Loustaunau (1994), which provides a canonical form for a system of multi-variable polynomial equations, such as (6.1) with rational coefficients. This enables us to establish when two systems are the same. Furthermore, as described in Badger and Sangwin (2011), these techniques can also be used to establish when:

- The student's system is inconsistent and so its variety is the empty set.
- The student's system is under-determined; its variety contains the teacher's.
- The student's system is over-determined; at least one equation should not be there.

We can also establish which equations in the student's answer are the cause of the above problems.

Inequalities are a very interesting, but somewhat neglected, area in CAS. A single inequality can be written in a canonical form, $ex \geq 0$ or $ex > 0$. There is some scope for teacher choice in what notion of 'same' is used: is $2x > 2$ acceptable? Systems of inequalities are much more interesting, since they are used to describe subsets of the real line. Many CAS currently have surprisingly poor libraries for manipulating real intervals and systems of inequalities, even in one variable. It is therefore unsurprising that few mechanisms exist to establish if two systems of inequalities represent the same situation.

6.9 Other mathematical properties we might seek to establish

Other useful tests include the ability to confirm that ex is continuous or differentiable in the variable v at the point v=vp, which both rely on the notion of a limit. We do not comment on whether such tests are computable in the formal sense.

For science, we need to deal with units; for example, 3.0 ms^{-1}. Such a test confirms

1. that the numerical information is correct, and of the required accuracy;
2. that the units are correct.

We have a number of issues to resolve in the case when this test fails. Are units of any kind supplied? If so, do these match those of the teacher? If not, are they dimensionally consistent with those of the teacher? (e.g. $km\,h^{-1}$ vs $m\,s^{-1}$?). In the case of consistent units should we convert the numerical information supplied? If no units are supplied, can we find a change of units in which the student's answer is 'correct'? Or, if the change of units needed to derive the correct number are dimensionally inconsistent, what feedback should we provide? A detailed discussion of these issues is outside the scope of this book, but it is sufficient to say that the problem is non-trivial.

6.10 Buggy rules

Another very useful facility is the ability to implement in CAS so-called 'buggy rules'. These are algebraic moves which are not valid. For example, the rewrite rule $(a+b)^n \to a^n + b^n$ could be viewed as an over-generalization of the distributive law. Further examples are adapted from Matz (1982) in the list below, though there are many others.

- Evaluation as juxtaposition: If $x = 6$, $4x \to 46$.
- Minus sign difficulties: $x = -3, y = -5$, $xy \to -8$.
- Incorrect parsing: $2(-3) \to -1$, $(-1)^3 \to -3$.
- Identity difficulties: $a\frac{1}{a} \to 0$, $0 \times a \to a$.
- Distribution/linearity: $(a+b)^n \to a^n + b^n$, $\sqrt{a+b} \to \sqrt{a} + \sqrt{b}$, $\sin(a+b) \to \sin(a) + \sin(b)$.
- Partial distribution: $2(x+3) \to 2x + 3$, $-(3a+b) \to -3a + b$.
- Excessive distribution: $a \times (bc) \to ab \times ac$.
- Fraction difficulties: $\frac{a}{b+c} \leftrightarrow \frac{a}{b} + \frac{a}{c}$, $\frac{a+b}{c+d} \leftrightarrow \frac{a}{c} + \frac{b}{d}$.
- Laws of indices: $2^{a+b} \to 2^a + 2^b$, $2^{ab} \to 2^a 2^b$.

Teachers will, no doubt, recognize many of these from students' work. For subtraction of multi-digit integers alone, Burton (1982) found more than sixty sub-skills which form part of the correct algorithm. Their *skill lattice for subtraction* details the relationships between the skills they identified. It is a salutatory reminder of the genuine complexity of the classic elementary algorithms. This is reproduced in Figure 8.11.

Rewriting rules, both valid and buggy, are an important topic in computer science and CAS design. We are not suggesting that students learn rules as a computer might, though such rules form the basis of some theories of learning. Clearly, a single response is no evidence of a deeply held misconception. It could easily be a slip or momentary lapse. Most of us make plenty of these. Tracking the use of these rules by students enables a profile of student knowledge—a *user model* to be built up—and this can form part of an adaptive testing or tutoring system. This is precisely the goal of expert systems, and the artificial intelligence approach to testing. We shall return to this issue in Section 8.6. Without the ability to establish that a student's work is consistent with certain kinds of buggy rules, profiling and adaptive testing are impossible. Suitable CAS tools are precisely those needed to enable it.

A computer algebra system such as Maxima in which all simplification can be suppressed, and with sufficiently fine control over the elementary operations, can be used to implement each of these buggy rules. The system can then take an expression, for example $(x+2)^3$, which a student is asked to expand, and generate the expression $x^3 + 2^3$, using one of a suite of buggy rules. It could also apply sequences of rules, mixed between genuine and buggy.

In Section 1.1 we discussed multiple-choice questions. When authoring a multiple-choice question the teacher must use their knowledge of likely incorrect answers to construct a list of plausible but incorrect responses. These *distracters* can often be conceptualized as the result of applying buggy rules. If we randomly generate versions of a question, then we need to operate on mathematical expressions in a buggy way to generate these. Even when we do not need to generate explicit distracters to show the student, if the teacher knows what a student is most likely to get wrong, then we can still check for this. If we can establish that the student's answer is consistent with the operation on the question with a buggy rule or sequence of rules, some of which are buggy, then we might choose to give feedback such as the following:

> Your answer appears to be consistent with. . . . Is that how you got your answer?

Unfortunately, getting a correct answer might be consistent with a buggy-rule. For example, we might *cancel like numbers* in the following fractions:

$$\frac{16}{64}, \quad \text{or} \quad \frac{19}{95}.$$

If a student 'moves' $3(x-2)$ to the left-hand side of $6x - 12 = 3(x-2)$ but forgets to change the sign, then they have made a mistake: $6x - 12 + 3(x-2) = 0$. In this case the answer still turns out to be correct.

The formal manipulation

$$\frac{p_1}{q_1} + \frac{p_2}{q_2} = \frac{p_1 + p_2}{q_1 + q_2},$$

sometimes called naive addition of fractions, is not always a 'buggy rule'. Given two rational expressions $\frac{p_1}{q_1}$ and $\frac{p_2}{q_2}$, the quantity $\frac{p_1+p_2}{q_1+q_2}$ is known as the *mediant*. It has uses in dealing with Farey sequences, for example, Hardy and Wright (1960).

An answer might be consistent with more than one bug. For example, an answer to $\int e^x dx$ which is missing the constant of integration could look like differentiation. Hence, there are cases when the culprit 'bug' cannot be identified with any certainty. Therefore, if we are looking for evidence of 'buggy understanding' then we need to generate questions carefully, to avoid false inferences or giving ambiguous feedback. Questions can also be tested automatically to decide if the correct answers to particular versions potentially result from buggy sequences of rules.

6.11 Generating outcomes useful for CAA

So far we have concentrated on predicate functions which return either true or false, to indicate whether the expression has a particular property. When teaching, it is not just that the student's answer fails to have a particular property which matters, but why it fails. To help students we would like to provide *outcomes*. Therefore, in practical situations it is invaluable to have a higher-level 'answer test' which returns the following:

1. **Errors**
   Error trapping is vital, including mathematical problems such as 'division by zero'.

2. **Validity**
   'Validity' is different from 'correctness'. For example, it would be invalid to seek to establish the algebraic equivalence of an equation with a set.

3. **Result**
   Either *true* or *false*. We have chosen not to return a number between 0 and 1 as partial credit since there is unlikely to be consensus on what should be awarded in particular cases. It is probably more sensible for any partial credit to be handled at the level of the question. That is, the teacher should assign partial credit on the basis of this result for each individual question. For example, we might be establishing equivalence with an expression derived from a common mistake, and while the outcome would be *true* it is unlikely there would be any marks.

4. **Feedback**
   In the case of a *false* result it is very helpful to have feedback automatically generated by the answer test. An example of this kind of feedback is given in Figure 6.2. Whether it is appropriate for this to be displayed to the student depends on the context in which the test is applied. It would be wasteful (and potentially error-prone) to have a separate mechanism to generate this feedback at a later stage.

5. **Answer note**
   The feedback may contain a manipulation of the arguments to the test, e.g. calculations involving the student's answer. Since questions are often generated randomly, and correct answers are not unique, for the purposes of statistical analysis the feedback is useless. The answer note string enables the outcomes of the test to be encoded at a finer-grained level of detail than that of the result field.

As a result of the requirement for feedback and a note, there is an asymmetry between the arguments in an answer–test which might appear symmetrical in a practical implementation. For example, in establishing algebraic equivalence between two expressions there is no particular order to the arguments. However, when providing feedback there is an asymmetry: the teacher's answer is assumed correct and we test the student's against this. Hence, we assume that the first answer to a test is nominally the student's answer and the second argument is the teacher's answer. The order is of course only relevant when the feedback generated by the answer test is to be shown to the student.

6.12 Side conditions and logic

Many of the problems with existing CAS arise from the way assumptions and side conditions are tracked, or ignored. For example, few CAS track the condition $k \neq -1$ when calculating $\int x^k \, dx$. When asked to perform this integral, the answer $\frac{x^{k+1}}{k+1} + c$ is returned with no restrictions on the value, and k can subsequently be assigned. Many similar examples are given by Stoutemyer (1991).

Many CAS have mechanisms for the user to state assumptions, e.g. declaring that n is an integer, or x is a positive real number. However, very few systems track any side conditions which arise in the course of a computation as a result of legitimate operations. A stark example is possible in Maxima. Take the equation $a = 0$ and divide both sides by a. This is achieved with the command (a=0)/a;. Maxima will return $1 = 0$, which is nonsense. Really, Maxima could have tracked the condition $a \neq 0$ when we divided by a. The fact that $a \neq 0$ and $a = 0$ from the original equation are a contradiction, and so the system should throw an error alerting the user to this problem. Many interesting 'paradoxes' can be generated from mistakes of this kind, e.g. Maxwell (1959) is a classic collection.

One solution to this problem is suggested by Beeson (1989), and implemented in the MathExpert system we describe in Section 8.10. During a calculation, the system builds up a set of side conditions, such as $k \neq 0$ or $x > 0$.

When an operator is applicable, but has a side condition, we proceed as follows:

1. First attempt to infer the side condition. [...]
2. If this fails, then attempt to refute the condition [...]
3. If this too fails, then *assume* the condition.

(Beeson, 1989, p. 210)

An example is shown in Figure 8.15 in which the expression $\sqrt{9x - 20} < x$ has the implicit assumption that $9x - 20 \geq 0$. Later, when we generate the condition $x < 4$ or $5 < x$ we need to keep track of the original condition to provide the correct final answer $\frac{20}{9} \leq x < 4$ or $5 < x$. This is done automatically, reducing (indeed eliminating) errors in the calculations.

Notice that to do this we need to work with systems of inequalities and sets of real numbers. We also need to combine calculations, inequalities, and the logical operations 'or' and 'and'. Many CAS have very poor libraries for doing this, even in comparatively simple cases which are perfectly computable.

Lastly, we note the impossibility of divorcing calculations and logic. Two example rules should serve to illustrate this.

$$\text{If } a \times b = 0 \text{ then } a = 0 \text{ or } b = 0.$$
$$\text{If } a^2 = b^2 \text{ then } a = b \text{ or } a = -b.$$

We could rephrase these in terms of set theory, e.g. the first of these might return a set $\{a = 0, b = 0\}$, but this means the same thing and simply obscures the central logical issue. Very few CAS are able to combine logic and theorem proving components together with side conditions and calculations. Despite the impressive calculation abilities of CAS, much work still needs to be done in this area to provide a reliable useful tool, even for elementary mathematics.

6.13 Conclusion

This chapter has considered, at a rather technical level, how to establish the properties of individual mathematical expressions. The desire to automate elementary algebra reveals some of the ambiguities and complexities of this subject. A practical teacher should not need to concern themselves with the technical implementation details on a day-to-day basis. The CAA designer should provide the question author with a suite of well-developed and reliable tools with which to assess answers to questions. However, the teacher does need to articulate the properties they seek in a way which differs from that of the traditional paper-based environment. Since it is rare that only one property is sought, there must also be mechanisms for applying a number of individual tests and on the basis of these assembling outcomes.

7

The STACK CAA system

In Chapter 2 we provided a vignette of automatic computer aided assessment of mathematics taken from STACK, a *System for Teaching and Assessment using a Computer algebra Kernel*. This relies on a computer algebra system to support mathematical tasks, such as representing and manipulating mathematical expressions. The vignette in Chapter 2 provides an introduction to how students interact with STACK questions, and the design of the formative feedback. In this chapter we describe this system in more detail. We also provide a more detailed case-study of use. Specific details of the current version are available online. While the purpose of this chapter is not a software manual some technical discussion is necessary.

7.1 Background: the AiM CAA system

AiM, *Assessment in Mathematics*, is a web-based CAA system for mathematics. AiM was originally written by Norbert Van den Bergh and Theodore Kolokolnikov at the University of Gent in Belgium, and was released in 1999; Klai *et al.* (2000). Subsequent developments were written by Neil Strickland at the University of Sheffield in the UK, together with contributions from a number of others; see Strickland (2002). AiM continues to be used, predominantly by university mathematics and engineering departments.

AiM is a collection of Maple worksheets which drive an interactive website. It is a stand-alone Internet-based assessment system which implements a hierarchy of subjects, quizzes, and individual questions. Students must use Maple's typed linear syntax to answer questions, and only the final answer is assessed. Teachers can author their own questions using AiM's scripting language and multi-part questions are possible. An example is shown in Figure 7.1.

It was the practical experience with AiM combined with a desire to improve the experience of students who use CAA which motivated the development of STACK. Starting in 2004 at the University of Birmingham, in collaboration with Laura Naismith and the University *Centre for Educational Technology and Distance Learning*, STACK was designed and developed by the author. Code for version 2.0 was developed with Jonathan Hart,

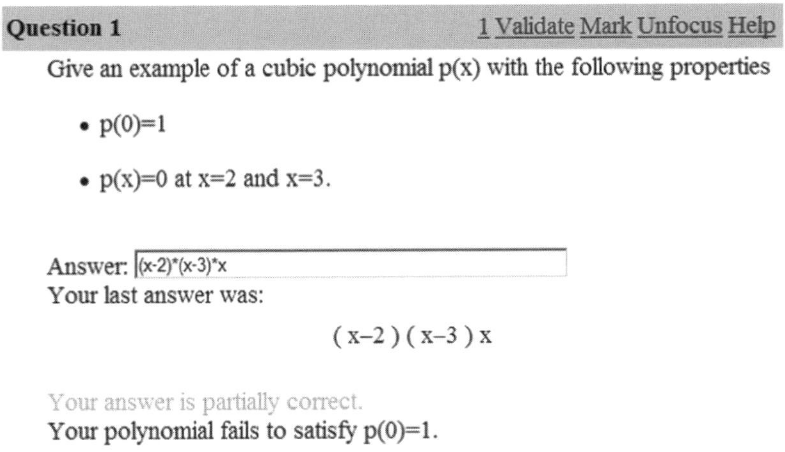

Question 1   1 Validate Mark Unfocus Help

Give an example of a cubic polynomial p(x) with the following properties

- p(0)=1
- p(x)=0 at x=2 and x=3.

Answer: (x-2)*(x-3)*x
Your last answer was:

$$(x-2)(x-3)x$$

Your answer is partially correct.
Your polynomial fails to satisfy p(0)=1.

Figure 7.1: An AiM question, as seen by a student.

and code for version 2.1 with Simon Hammond. Substantial support was provided by the Higher Education Academy's *Maths Stats and OR Network*. During 2011–12, version 3 was written in collaboration with Tim Hunt at the Open University, with contributions from Matti Harula at Alto University, Helsinki, and Matti Pauna at the University of Helsinki. Numerous other colleagues have contributed to the design, code, testing, and documentation. STACK is regularly used by a number of universities and colleges, see Sangwin (2010), and the code base is sufficiently reliable for regular use with very large groups of students. STACK is available in English and has been translated into Finnish, Swedish, Portuguese and Japanese (see Nakamura (2010)).

7.2 Design goals for STACK

The goal of STACK is to provide a useful and reliable assessment system for mathematics in which a student enters a mathematical answer in the form of an algebraic expression. The system should then establish the properties of this expression and provide outcomes in the form of feedback, a numerical mark, and an internal 'note' for later analysis. We chose an Internet-based system rather than a networked desk-top application. Therefore the interactions are based on the client–server model through a web browser. This dictates many of the later decisions, such as submitting a page of answers and with it the validate/mark protocol.

- STACK generates random versions of questions in a structured mathematical way;
- accepts answers from students which contain mathematical content, rather than MCQs;
- establishes the mathematical properties of those answers;

- generates outcomes which fulfil the purposes of formative, summative, and evaluative assessment;
- stores data on all attempts at one question, or by one student, for analysis by the teacher.

Once we have established mathematical properties of answers, for formative assessment at least, we would like to provide feedback to the student almost instantaneously. It was an important goal that this feedback should be able to contain the results of calculations which include the student's answer, such as that shown in Figure 2.3. It was felt insufficient to restrict the teacher to the use of static responses written in advance. The CAS is available, so it is a natural desire to make full use of it in this way. Surprisingly perhaps, not all systems which make use of a CAS enable the teacher to encode feedback.

We wanted to enable teachers to write their own questions. There is great variety in the extent to which teachers are able to write their own questions in CAA. In some systems the items are fixed. The teacher has no facilities to alter questions, or the context in which the questions are seen. The advantage of this approach is *quality assurance* of the items. In other systems the teacher is free to select questions from a list and assemble these into quizzes. In Web Interactive Mathematics System (WIMS) of Xiao (2001) the teacher is able to choose parameters within particular questions, but authoring questions is a task for an expert developer. With STACK the teacher can author questions from scratch, and has full control over the outcomes.

There are always tensions between (i) the freedom to change and adapt; (ii) quality control and reliability; (iii) simplicity and efficiency; (iv) the level of expertise needed. Actually writing questions for CAA is a rather difficult task, requiring expertise in mathematics, sensitivity to the needs of students and technical skill in the details of the CAA system. There are further tensions between (i) the educational objectives one might, ideally, try to encode; (ii) what it is possible to establish automatically; and (iii) what is achievable within the constraints of time and expertise available locally.

To make use of AiM teachers must write substantial pieces of Maple code to evaluate students' answers. STACK seeks to reduce this requirement, asking teachers to focus on *properties* such as algebraic equivalence, rather than on writing code such as

```
if simplify(SA-TA) then mark:=1 else mark:=0.
```

However, while the design of the marking algorithm in STACK encourages the teacher to articulate the properties they seek to establish, to provide sophisticated feedback the teacher still sometimes needs to write sophisticated CAS code themselves. While the attention of the CAA community is predominantly, and quite rightly, focused on the needs of the students, the needs of teachers as learners of the art of CAA are rarely, if ever, considered explicitly. In Sangwin and Grove (2006) teachers were described as *neglected learners*.

An explicit design goal was to provide a system in which every component is available under an open source licence. Our philosophy is to encourage people to collaborate on the core CAA system, the infrastructure, leaving them free and responsible for the way they use it in their own teaching. This is not anti-commercial, but it does require a business model other than seeking revenue for selling the software itself. A consequence is the relative ease with which it is possible to collaborate with other teams of like-minded software developers and teachers. The goals outlined above are too ambitious for a single individual or even a single large team. Therefore, we decided to make use of, and contribute to, related projects rather than develop everything for ourselves.

As a result of these goals the following decisions were made when designing STACK.

- **Content and identity management**
  Management of identities, e.g. usernames/passwords and roles, is a complex project in itself and there are many systems which do this very well. Currently, STACK provides a question type for the Moodle quiz module, Wild (2009), which in turn sits on multiple content and identification management facilities provided by Moodle.

- **Computer algebra**
  CAS are large software projects in their own right. STACK relies on the computer algebra system Maxima. When choosing a CAS, the goal of a truly open source project narrowed the field considerably, and Maxima was selected mostly on the grounds of its licence. Interestingly and fortunately, as discussed in Chapter 6, the design decisions made in Maxima make it particularly suitable for CAA.

- **Authoring mathematical text**
  The practical teacher needs to write significant fragments of mathematical text. The question, feedback, and worked solution all contain written mathematics. LaTeX is essentially the standard mark-up language used to typeset mathematical text. While there is a steep initial learning curve, it is particularly comfortable for experts to type. Some other standards, such as MathML or OpenMath, cannot be authored by hand and so require the use of an editor. We note that LaTeX does not encode the meaning of mathematical expressions but operates only at the typographical level. Hence, we need another format for encoding the meaning. The syntax of Maxima was a natural choice here. We also note the desire to include dynamic content, e.g. a GeoGebra applet. It is particularly important to dynamically generate mathematical expressions, images, and dynamic content, and insert these into the text, both while randomly generating questions and when producing feedback.

- **Display of mathematical text**
  Somewhat surprisingly there is still no standard way for all mainstream web browsers to reliably display mathematics, Hayes (2009). It is not the responsibility of the STACK project to solve this problem! However, STACK needs some way of converting the LaTeX code provided by the question author into a readable expression on a student's web browser. Moodle provides a *filter mechanism* and STACK relies on this, making no effort to convert the LaTeX code within questions into something else. Currently, the preferred filter is MathJax.

Since STACK questions form part of a Moodle quiz it is perfectly possible for the teacher to include other questions alongside STACK questions. This fixed linear quiz structure, in which students can choose to answer the questions in any order, is only one possible model for CAA. Adaptive testing, in which the next question is determined by the outcome of the current one, is another model we examine later. Implicit here are decisions made by the designers and implementors of Moodle. This affects access to repeated attempts at quizzes and also the availability of such things as model solutions after any quiz due date.

In developing STACK we somewhat naively assumed that there would be a clean separation of the 'question' from the 'quiz'. That is, it would be possible to 'insert questions' into a more general quiz structure in a flexible way. STACK 2 inserted questions into Moodle quizzes, without really being part of Moodle itself. We could not implement this in a satisfactory way, which was a significant and unexpected surprise. For example, it is common practice to ask a group of students to complete a 'quiz'. During this, formative feedback is available, and where necessary multiple attempts are encouraged to help students ultimately succeed in these tasks. However, there is an end date and time, after which the teacher's model solutions become available and further attempts by the student are prevented. The concept of 'due date' is a property of the 'quiz', but the 'worked solution' belongs to the 'question'. The question needs to be able to access data fields within the quiz to behave correctly, and so a clean separation of data structures is impossible. Moodle now has an abstraction layer between a quiz and question called *question behaviours* which facilitates this, and as a result STACK 3 is much more tightly integrated into Moodle.

7.3 STACK questions

STACK has developed a *data structure* with which we may represent a large range of mathematical questions together with an *interaction model* through which these can be used. The core questions we sought to automate support the direct instruction of technique associated with classic mathematical algorithms. This includes algebra, calculus, linear algebra, differential equations, and transform theory. The data structure used by STACK contains a number of separate fields.

The *question variables* are CAS statements which are evaluated to create a version of the question, including any randomization. Both the content form and the displayed form of each variable are stored for later use in other parts of the question. The *question stem* is the text shown to the student. This is, literally, the 'question'. STACK inserts the displayed form of any randomly generated expressions, and also the *interaction elements* through which the student responds to the question. For example, in Figure 2.1 the expression $3x^5$ is randomly generated and inserted into the integral. The form box is the single interaction element. The *potential response trees* are algorithms which take the student's answer(s) and establish the mathematical properties. I.e. this is the code which assesses the question, and generates outcomes, such as feedback, a score (mark), and notes for statistical analysis. The optional *general feedback*, also called a worked solution, is text which is shown to the student *after* the due date for a quiz. Unlike genuine feedback, a design decision was made that the worked solution may not depend on the student's answers. A *question note* is used by the teacher to leave an intelligent 'note to self' about the specific values of question variables which a student has been given. Two versions of a question are considered to be identical if they have identical question notes, so an empty note is unhelpful. Random questions must have such a note which is used to establish that each random version is valid, and to establish the relative difficulty.

Fulfilling the reliability goal for large groups of students has delayed the implementation of features such as (i) assessing steps in working, (ii) free input for online tutoring (compared with assessment of only the final answer), and (iii) adaptive testing. These are topics to which we shall return in later chapters when we compare STACK with the work of others in Chapters 8 and 9.

7.4 The design of STACK's multi-part tasks

Chapter 2 described a minimal single-part question. Many teachers will be satisfied with single-part questions, although our experience of CAA is that the availability of multi-part tasks reduces the distortion of questions and enables the teacher to encourage a particular method. Compare the following two examples.

▼ **Example question 14**

Solve $x^2 - 10x + 21 = 0$.

▼ **Example question 15**

Factor $x^2 - 10x + 21$. Hence, solve $x^2 - 10x + 21 = 0$.

In the second the use of two parts strongly suggests that a particular technique should be used. A multi-part question, with an interaction element for each part, can be used to assess the second. While this clearly stops short of accepting an arbitrary mathematical argument, such multi-part items enable a wider range of valid assessments to be implemented. Our discussion of multi-part items is via a collection of examples. First we consider the following, rather elementary, task.

$$\text{Expand } (x-2)(x-3). \tag{7.1}$$

For the sake of argument it has been decided that it is educationally appropriate to provide students with the opportunity to 'fill in the blanks', as shown in Figure 7.2. The student thus sees two independent form boxes, and may choose to complete one at a time. The teacher has decided that the student must complete all boxes before any feedback is assigned, even if separate feedback is generated for each interaction (i.e. coefficient).

Consider the sequence of four tasks shown in Figure 7.3. Mathematical properties such as the *odd part* of an expression, i.e. $\frac{1}{2}(f(x) - f(-x))$, can readily be computed by the CAS. A function is even if it has no odd part. On the basis of properties such as this, outcomes are assigned. In the example shown, the student has misinterpreted what it means for a function to be 'odd and even', and has entered an expression which is only partially correct. Each of the four interaction elements has an independent algorithm which establishes the desired

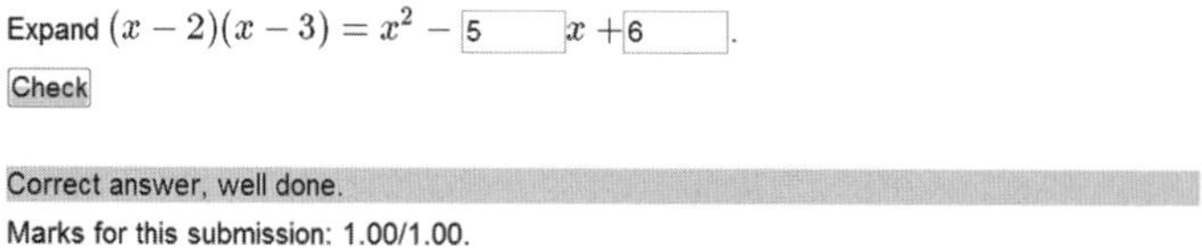

> Expand $(x-2)(x-3) = x^2 -$ [5] $x +$ [6] .
> [Check]
>
> Correct answer, well done.
> Marks for this submission: 1.00/1.00.

Figure 7.2: Many parts, one assessment algorithm.

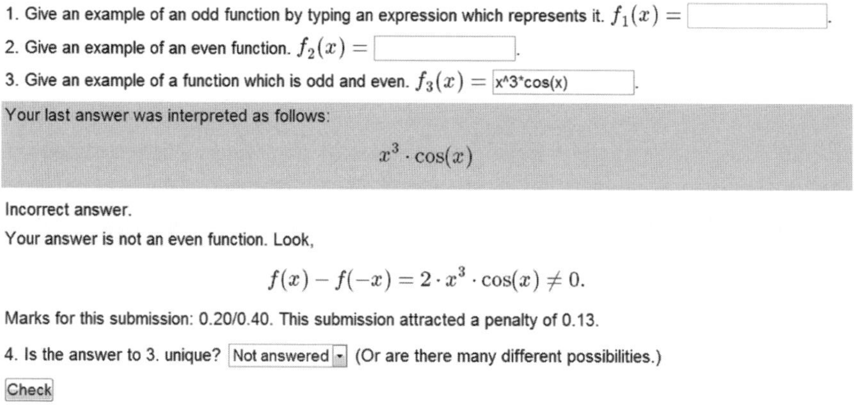

> 1. Give an example of an odd function by typing an expression which represents it. $f_1(x) =$ [ ].
> 2. Give an example of an even function. $f_2(x) =$ [ ].
> 3. Give an example of a function which is odd and even. $f_3(x) =$ [x^3*cos(x)] .
> Your last answer was interpreted as follows:
> $$x^3 \cdot \cos(x)$$
> Incorrect answer.
> Your answer is not an even function. Look,
> $$f(x) - f(-x) = 2 \cdot x^3 \cdot \cos(x) \neq 0.$$
> Marks for this submission: 0.20/0.40. This submission attracted a penalty of 0.13.
> 4. Is the answer to 3. unique? [Not answered ▾] (Or are there many different possibilities.)
> [Check]

Figure 7.3: Many parts, independent algorithms.

properties of the student's answer and assigns separate outcomes. Notice that the student does not need to answer all the parts at once, as they can be attempted and changed in any order. Notice also that the feedback in Figure 7.4 contains the result of a CAS calculation of the student's answer.

These two situations attempt to illustrate two extreme positions.

1. All interactions within a single multi-part item must be completed before the item can be assessed.

2. Each interaction within a multi-part item can be assessed independently.

Devising a data structure to represent multi-part questions which satisfy these two extreme positions would be relatively straightforward. However, it is more common to have multi-part questions which are between these extremes.

The question shown in Figure 7.4 is a typical example. The first part asks the student to write an equation which models the situation. In fact, if we take x to be the length of the shortest side then one equation which captures the statement is $x(x + 8) = 48$. Notice that the teacher has chosen not to specify the name of the variable that the student should use, or which side this variable represents. It has been decided that a statement such as 'use x as the length of the shortest side' would give far too much away, and would render the question significantly easier, potentially reducing its validity. Given this flexibility, care is needed when automatically establishing the properties sought. We want an equation from

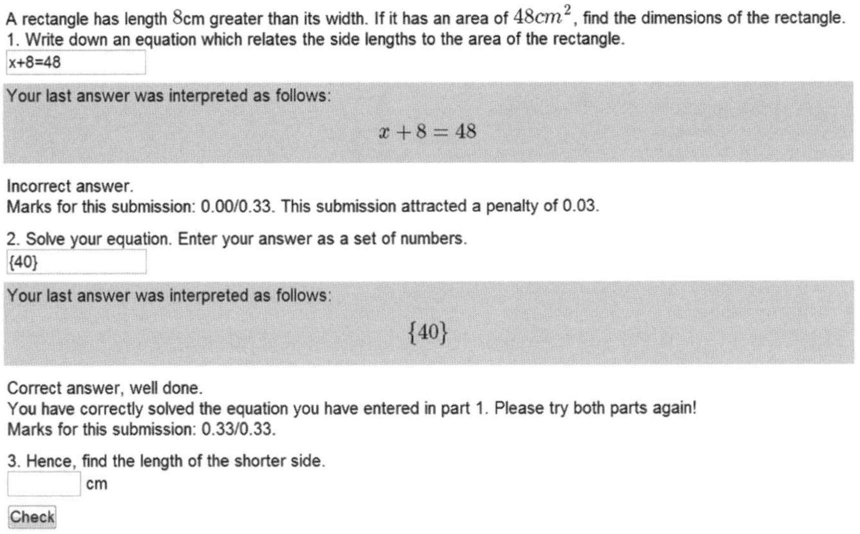

A rectangle has length 8cm greater than its width. If it has an area of $48cm^2$, find the dimensions of the rectangle.
1. Write down an equation which relates the side lengths to the area of the rectangle.
[x+8=48]

Your last answer was interpreted as follows:

$$x + 8 = 48$$

Incorrect answer.
Marks for this submission: 0.00/0.33. This submission attracted a penalty of 0.03.

2. Solve your equation. Enter your answer as a set of numbers.
[{40}]

Your last answer was interpreted as follows:

$$\{40\}$$

Correct answer, well done.
You have correctly solved the equation you have entered in part 1. Please try both parts again!
Marks for this submission: 0.33/0.33.

3. Hence, find the length of the shorter side.
[   ] cm

[Check]

Figure 7.4: Many parts, many algorithms.

the student which is equivalent to either $x(x + 8) = 48$ or $x(x - 8) = 48$, regardless of which variable the student has chosen. That is to say, if the student's equation has a single variable we substitute x for this before we make the comparison.

The answer shown in Figure 7.4 to the first part is clearly wrong. Furthermore, the student has also answered the second part. Asking the student to enter the solutions as a set does not tell the student how many solutions to expect: no solutions is also possible. Providing two answer boxes would be too obvious, and might provide the student with a hint that a quadratic is needed for the first part. Of course, we have assumed some familiarity with entering sets of numbers. Regardless of whether the student has the first part correct, we have chosen to consider whether they have actually correctly solved the equation *they gave for the first part*. This is known as follow-through marking. In order to assess the answer to the second part, the first interaction element must also be entered. The student's answer shown in Figure 7.4 is the correct solution, given an incorrect first part. The third, unattempted, part, has been implemented to be independent of the other two. In this example the student could type in a number without answering the previous two parts.

We could argue about whether the preceding choices have any educational merit. The example is included here to illustrate the dependencies between interaction elements (IE) and algorithms (Alg), which may be visualized as follows.

	IE		
	1	2	3
Alg. 1	●		
Alg. 2	●	●	
Alg. 3			●

In general, many patterns are possible for linking interaction elements to algorithms. In the example shown in Figure 7.2 we had one algorithm and two interaction elements. Here, all the interaction elements are required before assessment can take place. The dependency is simply as follows.

	IE	
	1	2
Alg. 1	•	•

A 'leading diagonal' would indicate separate items, with one algorithm for each. The questions in Figure 4.1 would be assessed in this way, as would those of Figure 7.3.

	IE			
	1	2	3	4
Alg. 1	•			
Alg. 2		•		
Alg. 3			•	
Alg. 4				•

Often the dependencies naturally fall between these two extremes, such as that of Figure 7.4.

Where follow-through marking takes place we visualized this matrix as a 'lower triangular' block. This can also be used to assess steps in a calculation, where it is thought appropriate to structure the assessment into pre-defined steps. In such a block, the first algorithm uses only the first interaction element. The second algorithm uses the first and second interaction elements, and so on. Arbitrary configurations are possible, though it would be relatively easy to create scenarios which make little sense to the student. However, it is possible to create some situations which seem strange at first but which are really quite natural. For example, an interaction element with no corresponding algorithm is also known as a *survey item*. If there are a number of separate properties to establish, it may in some circumstances make more sense to have one algorithm to establish each of these and provide outcomes separately. This could be combined into one algorithm, but need not be.

STACK provides a data structure which permits all these variations. The crucial observation is a complete, but somewhat counterintuitive, separation between two important components:

1. A list of *interaction elements*.
2. A list of *assessment algorithms*, known as *potential response trees*.

The design is based on a dichotomy between the student's and teacher's view point. To the student, a question is multi-part if it has more than one *interaction element*. These are those things with which the student really interacts. The prototype is a mathematical expression entered in a form box using a typed linear syntax. To the teacher a question is multi-part if it can be broken down into sections, each of which is assigned separate *outcomes*. At least one interaction element is required for each algorithm to be executed, and all interaction elements upon which an algorithm relies must be syntactically valid. Each interaction element may be used by a number of separate algorithms. Indeed, an interaction element may be used by no algorithm: the result is a survey which is not assessed. These parts all sit in the context of random variables, which determine the smallest sensible unit of technical item encoding. It is relatively common to ask a number of separate questions about a single randomly generated object.

7.5 Interaction elements

The question stem, i.e. the text actually displayed to the student, may have an arbitrary number of *interaction elements*. The default is a form box into which the student types a mathematical expression. An element may be positioned anywhere within the question stem. There are a number of options for these interaction elements.

1. The name of a CAS *answer variable* to which the student's answer (if any) is assigned during response processing, e.g. ans1, ans2, etc. The teacher uses these variable names to refer to the student's answer in calculations, within the potential response tree and in any feedback.
2. The *type* of the interaction element, such as the following:
    (a) An input box for, *direct linear algebraic input*, e.g. 2*e^x.
    (b) A graphical input tool, e.g. DragMath.
    (c) A matrix grid of specified size.
    (d) True/false selection.
    (e) Multiple choice, dropdown list, etc. Despite our comments in Section 1.1, STACK does support multiple-choice interactions.

    The only essential requirement is that the internal result is a valid CAS expression.
3. The teacher's correct answer.
   It was a design decision to require the teacher to specify one 'correct' expression for each interaction element.

Let us assume that a version of a question has been created. The student will interact with some of the elements in the question, and once they are ready they will ask for the item to be assessed. In CAA based upon traditional HTML pages containing forms, submitting a page is a well-defined event. We term this an *attempt*.

Syntax errors are a practical problem, and we did not want to penalize answers on the basis of a technicality. Therefore, STACK implements a two-step assessment mechanism which separates out

1. validation, from
2. assessment.

Since validation is a property of the expression, this is tied to the interaction element and makes it (mostly) independent of the context of the question, or the potential response tree.

The central difficulty is to decide what action should result when a student changes an existing answer to more than one interaction element in a multi-part item, but then submits the page. To keep track of the logic, each interaction element has an internal *status*. For example, `blank` indicates that the interaction element has not been previously given a value by the student, or the field is now empty since the student has deleted an answer. The other states include `invalid`, `valid` (but not yet assessed) and `score`. In the last case the answer is available to any potential response tree requiring it.

In fact, the process of validation includes much more than a syntax check. In practice there is a range of reasons why an expression might be rejected as invalid.

1. If an informal syntax is acceptable, the student's answer should be modified to be correct CAS syntax. e.g., insert `*`s where necessary, or remove white space, e.g. change `sin x` to `sin(x)`.
2. The syntactic validity of the expression is checked.
3. Check for any forbidden strings particular to this question, e.g. `diff` commands.

If the status at this stage is `valid` then the student's answer is evaluated by the CAS. This implements further checks, some of which are optional.

1. Check for mathematical invalidity or errors which arise from *evaluation*, e.g. `1/0`.
2. Optional pedagogic validity is ensured.
   (a) Are floating point numbers acceptable?
   (b) Are all rational numbers required to be in lowest terms?
   (c) Do we require the student's expression to be the same type as the teacher's 'correct answer'? E.g. expression, equation, inequality, set, list, or matrix.

Notice that one option is to reject expressions which contain floating point numbers as *invalid* not 'wrong'. These pedagogic options have implications for how numerical marks are assigned for repeated attempts.

If the student's answer is valid, then the student's answer is displayed to them in a traditional two-dimensional notation. An example is shown in Figure 2.2. Once all the interaction elements upon which a particular potential response tree rely have status `score` then the actual assessment takes place.

7.6 Assessment

At last we come to the actual assessment of students' answers. The *potential response tree* is the name for the algorithm which assesses the student's answer and assigns outcomes. Before we can traverse the tree we need to set the randomly generated context which includes the *question variables*.

The tree is an acyclic directed graph consisting of *nodes*. In each node two expressions are compared with an answer test. Informally, the answer tests have the following syntax:

```
[Error, Result, Feedback, Note] = AnswerTest(SA,TA,opt)
```

The result, either true or false, will determine which branch of the node will be followed. The teacher specifies how numerical scores are adjusted for each branch, and any feedback is concatenated with previous feedback. The additional notes are stored to record the path taken through the tree during this attempt. Lastly, the next node is nominated or the process terminates. Note that at this point it would be possible to expand STACK to specify which question the student should see next for adaptive testing, though we have not implemented this yet.

STACK encourages students to make repeated attempts where necessary. By default, a small percentage (e.g. 10%) of the marks available for the question are deducted for each different valid but incorrect attempt. This is shown in Figure 2.6. However, the final mark is the maximum mark over all attempts, so a student is never worse off trying a partially correct question again. Other schemes for manipulating marks are available, including simply taking the mark for the last attempt.

The *answer note* provides a record of the result of applying each test and the subsequent route taken through the potential response tree. This is useful to the teacher for reporting statistics on achievement. Since students may be answering random versions we need an unambiguous record of their route through the tree, regardless of their answer or the random context. For example, we might be interested in whether a given random version has a disproportionate occurrence of particular notes. This might indicate a problem with one random version, which might be unfair in high stakes situations. Alternatively we might use the notes to examine if a student's answers to a sequence of questions is consistent with a 'buggy rule'; see Section 6.10. A single numerical mark, and the feedback strings shown to the students, are not adequate for this task. The answer notes also play a crucial part in quality control and testing, as described in Section 7.7 below.

The score, feedback, and notes correspond approximately to the summative, formative, and evaluative purposes of assessment respectively.

7.7 Quality control and exchange of questions

A practical teacher regularly faces two problems: (i) testing questions thoroughly, i.e. quality control; and (ii) understanding how a question written by someone else will behave, i.e. the feedback and numerical marks it will provide to a range of answers. STACK provides a single mechanism to help a teacher address both problems, akin to *unit tests* in software engineering.

When authoring a question, the teacher may define an arbitrary number of *question tests*. For each interaction element the teacher may specify a response, or leave this element blank to indicate the student did not yet respond to this part. This response can depend on any of the question variables to enable the 'answer' to reflect the random version the student really sees. For each potential response tree they then choose the intended outcome from a list of possible answer notes. Recall that the answer notes record the logical outcome of traversing the potential response tree, and hence provide sufficient information to make a fine-grained evaluation.

When testing a question, STACK takes each test in turn, automatically marks a fresh instance of a particular random version of the question, and then assesses these answers. It establishes whether the answer notes obtained match those nominated by the teacher when authoring. This reduces the need for the teacher to repeatedly 'play' with potentially wrong responses and a number of random versions. Furthermore, if the question author spots and fixes a bug with the potential response tree, or other code, they can easily check all these cases again automatically. If a teacher looks at someone else's question, they can immediately see how the system will respond to a range of potential answers with full details of the marks, feedback, and notes given. These tests are an integral part of the data structure, and so are carried around, stored, and exchanged with the item itself.

7.8 Extensions and development of the STACK system by Aalto

An extensively modified version of STACK version 1 has been used by the School of Science and Technology at Aalto University[1] since 2006, with substantial code written and maintained by Matti Harjula over a number of years. Aalto teach in three languages: Finnish, Swedish, and English, so they have translated the system into Finnish and Swedish. In addition to translation, these modifications include:

- Authentication of students through local institutional systems, see Harjula, 2008, Chapter 6.
- Enhanced support for, and display of, $\LaTeX$.
- 'Multi-part questions'.
- Different interaction types, e.g. GeoGebra worksheets.
- 'Question blocks'.
- Improved logging of student data.

STACK version 1 made use of a software package TtH to convert $\LaTeX$ into pure HTML. At that time, displaying mathematics satisfactorily online was still very difficult and there were significant problems with browser compatibility. This sometimes resulted in unreadable formulae. Clearly this is unacceptable, particularly in high-stakes situations. The response

---

[1]. Established in 2010, Aalto University was created as the merger of The Helsinki School of Economics, Helsinki University of Technology, and The University of Art and Design Helsinki, see http://www.aalto.fi/en/about/. STACK was used at the Helsinki University of Technology.

Funktio f määritellään paloittain seuraavasti:

$$f(x) = \begin{cases} 1 & x < -1 \\ p(x) & -1 \leq x \leq 1 \\ \sin(\pi \cdot x) & x > 1 \end{cases}$$

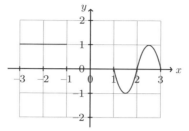

Määritä kolmannen asteen polynomi $p(x)$ siten, että funktio on jatkuvasti derivoituva. Funktiota kutsutaan jatkuvasti derivoituvaksi, kun sekä $f(x)$ että $f'(x)$ ovat jatkuvia funktioita.

Figure 7.5: An example of display from the Aalto STACK system.

of Aalto was to create image files containing the question content. This is highly robust, but not particularly accessible, e.g. users cannot scale fonts or use screen readers[2]. By taking this approach the Alto fork is able to take advantage of a wide range of LaTeX packages. For example, they support various diagrams such as that shown in Figure 7.5. Notice that this contains text, equation environments, and graphical images, combined into a single picture file.

STACK 1 provided support only for single input fields, and a single potential response tree. Aalto developed methods for multi-part questions, enabling a question to contain multiple input boxes. This significantly enhances the functionality of STACK. However, the more ambitious suggestions in Ruokokoski (2009) ultimately proved to be problematic in practice. Different interaction types are described in (Harjula, 2008, Chapter 7), and these include matrix grids with individual boxes for entries and HTML elements for multiple-choice list elements. The combination of multi-part questions and different input types provides a particularly rich and flexible platform through which to provide and assess mathematical questions. An example question is shown in Figure 7.6 which contains multiple input elements, and both formula entry into form boxes and 'radio button' choices. Assessment still relies on a single algorithm.

Inclusion of GeoGebra provides an alternative method of creating diagrams, but the configuration of the diagram may also constitute part of the answer. An example question, by Prof Gustaf Gripenberg, is shown in Figure 7.7. In this question the students must move points within the diagram as part of their response to the problem.

---

2. We note that STACK 2 abandoned TtH in favour of client-side Javascript libraries JSMath, and STACK 3 relies on the Moodle filtering system to render LaTeX equation environments, with MathJax being the strongly preferred option. In doing this STACK has moved away from support for other LaTeX features and packages, and closer to HTML with embedded LaTeX equation environments.

Olkoon annettuna graafi

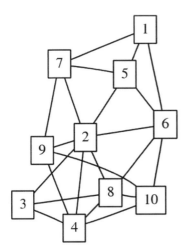

Onko graafille muodostettavissa Eulerin kierros tai polku? Merkitse saamasi polku tai kierros kuten [3,4,5,...,10]. Jos löysit kierroksen, laita alku- ja loppupiste molempiin päihin ketjua. Jos graafilla ei mielestäsi ole polkua tai kierrosta, jätä niiden esitykseen varattu kenttä tyhjäksi.Huomaa, että pääsääntöisesti sivua esittävät suorat viivat, teräviä kulmia niissä ei ole koskaan.

Figure 7.6: Muti-part question with randomly generated diagrams.

Question blocks extend the functionality of the basic text data-type. In STACK version 1 this was taken to be LaTeX into which CAS commands could be embedded. The question blocks enable sections of text, i.e. blocks, to be included or omitted, based on the values of CAS parameters. As a specific example we include the following code:

```
{% if @some_CAS_expression@ %}
   The expression seems to be true
{% else %}
   not quite true
{% end if %}
```

In the figure below three level curves of a function $f(x,y)$ are sketched. Move the point C so that the vector from A to C has the same direction as the gradient of f in the point A.

Move the point B as well so that the gradient in this point B has the same directions as the vector $V = [-.608643 \ -.793444]$.

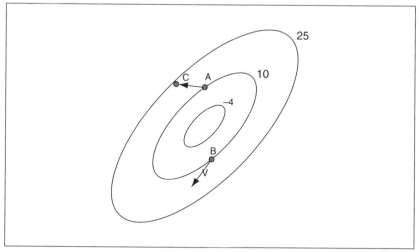

The vector from A to C is: [-0.796135122669] [0.078542131700]

The point B is: [0.20783657610] [-0.9123769354!]

Figure 7.7: A question including a GeoGebra applet.

The `some_CAS_expression` must evaluate to true or false, and on this basis one of the text blocks is included. This has primarily been used to typeset complex randomized diagrams, such as that shown in Figure 7.6. These blocks include the normal range of logical operations, including `else if` and `foreach-block` loops. A specific block environment sends its content out to special tools for processing, much in the same spirit that the CAS is called. We note that many of these features have been subsequently incorporated into STACK version 3.

7.9 Usage by Aalto

STACK was introduced to Aalto University by Dr. Antti Rasila in 2006 as part of the MatTaFi project, e.g. see http://matta.hut.fi/mattafi/. This substantial project was funded by the Finnish Ministry of Education as part of the *Finnish Virtual University*. See http://www.vy.fi/. After an initial pilot, Dr Jarmo Malinen joined the project, together implementing a new and more advanced experimental course. Many of these materials are

Table 7.1: Total student enrolments on courses using STACK, by year.

Year	Total
2006	148
2007	250
2008	430
2009	1197
2010	1481
2011	2189

still being used. Following this success, other colleagues became interested in STACK, and after the end of 2007 the department adopted the system. Since then a group has evolved to work with STACK.

Table 7.1 shows the total student enrolments for the period 2006–2011. During this period there was a steady rise of STACK usage from 148 students taking KP3-I in 2006, to 2,189 enrolments on thirteen courses by 2011. During this period nineteen courses have made use of STACK assessments. In the period of one academic year more than 67,000 question attempts were assessed automatically. Estimating that a person will take an average of one minute to mark a question, and an eight-hour working day, this represents 1,116 person-hours of work, or approximately 140 person-days.

To provide more information on how STACK is actually used, we consider one course in more detail. S1 is a *basic course in mathematics* and is the first of the three compulsory mathematics courses for electrical and telecommunications engineering students. The course contains material on complex numbers, matrix algebra, linear systems of equations, eigenvalues, differential and integral calculus for functions of one variable, introductory differential equations, and Laplace transforms. The intention is to provide a mathematical foundation for these degree programmes. A comparison between the initial *basic skills* test, traditional teaching and the STACK exercises was undertaken by Rasila *et al.* (2010). The main objective of that research was to '*measure the impact of e-assessment on learning outcomes in engineering mathematics*' (Rasila et al., 2010, p. 38).

An interesting feature of the use of STACK on this course at Aalto is the non-linear way by which marks for STACK coursework and examination are converted into a final grade. The most common method is, arguably, to sum coursework and the examination, then to convert this total into an overall course grade. To account for assessment uncertainty/errors, special attention is paid to students on borderlines to ensure that their grade reflects accurately their achievement. For example, at the University of Birmingham we typically set the coursework to contribute 20% and examination 80%, and then set thresholds for the four categories in the UK system, i.e. first class, upper second, lower second, and third at 70%, 60%, 50%, and 40% respectively, with overall marks below 40% being a 'fail'.

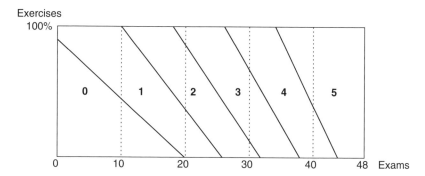

Figure 7.8: The grading system on the course discrete mathematics: proportion of exercises solved is on the *y*-axis and examination score (0–48 points) is on the *x*-axis. The grades are 0 (fail) and 1–5, where 1 is the least passing grade and 5 is the best. From Rasila *et al.* (2010).

Compare this with the grading system shown in Figure 7.8. The simple summation method would result in threshold lines having gradient minus one throughout. In Figure 7.8 the threshold lines are straight, but as the grade increases the lines have increasing negative gradient. This has the effect of increasing the importance of the exam for students seeking the higher grades. Notice that it is possible to pass the course on exercises only, i.e. without scoring on the exam.

However, there is a strong correlation between participation in STACK exercises and exam scores. In particular, Table 7.2 groups students by the final outcome of the course, as measured by a grade of 0 to 5, where 0 is a fail. The entries in the table show percentages of automatically assessed (above) and traditional (below) exercise assignments solved by students, and hence represents the level of coursework activity. It is clear that the general

Table 7.2: The percentage of automatically assessed (above) and traditional (below) exercise assignments solved by students. Numbers are sorted presented by the grade given (0–5), where 0 indicates failing the course. From Rasila *et al.* (2010), Table 2.

	0	1	2	3	4	5
2007	11.60	17.97	33.02	31.19	64.04	79.68
	3.78	7.77	20.19	9.40	26.84	61.61
2008	13.20	23.62	36.55	49.56	65.60	74.89
	4.79	13.56	16.15	28.85	56.81	58.44
2009	14.62	23.28	38.78	49.53	51.16	78.32
	3.77	10.00	29.20	50.48	68.22	92.48

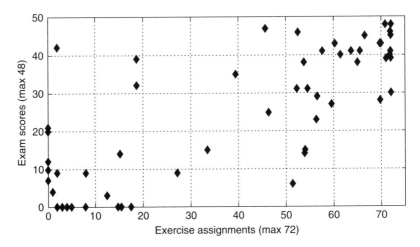

Figure 7.9: Student scores from examinations and exercises by the time of the mid-term examination. About 29% of students have solved more than 90% of exercises. From Rasila *et al.* (2010).

level of activity among the failing students is very low, and that the level of activity for the STACK quizzes is usually higher. Statistical analysis of the results from this course shows that the amount of time students spent with the system '*has a significant correlation to their scores from exams*', Rasila *et al.* (2010). This is further illustrated in Figure 7.9 which plots scores from exams and exercises by the time of the mid-term examination.

To address the problems of '*general passivity and lack of participation among the students*', Majander and Rasila (2011) considered a blended-learning approach. Each week students had 3 hours of face-to-face lectures after which six exercises were assigned. Two or more of the exercises were computer aided STACK exercises which were to be submitted via the Internet. The rest of the exercises were traditional pen-and-paper exercises, which could be solved either in the weekly face-to-face exercise sessions or handed in as written solutions. The total number of exercise assignments during the course was 72, and about 2/3 of these were STACK exercises. In addition, there were two voluntary examinations.

Using a modified e-Learning Experience Questionnaire, Majander and Rasila (2011) sought to understand how students perceived the quality of the course.

> In our final questionnaire, we had six categories concerning the quality of the course: quality of STACK exercises, clarity of goals and standards, appropriateness of assessment, appropriateness of workload, practical arrangements, and blended learning. (Majander and Rasila, 2011)

Students responded using a Likert scale where 1 = strongly disagree, 2 = disagree, 3 = neutral, 4 = agree and 5 = strongly agree. There was also a section for open comments, in Finnish. Thirty of the 58 students on the course responded to the questionnaire. The full anaylsis was provided by Majander and Rasila (2011). They conclude as follows.

> We have found out that using e-assessment as a method of continuous formative assessment is a flexible way to answer to some of the common issues in using exercise assignments as a part of the assessment procedure. The feedback concerning the quality of the course suggests that the new arrangements did not result in too heavy a workload on the students. (Majander and Rasila, 2011)

And in general, a positive if cautious conclusion.

> The learning outcomes as measured by the grades were much better compared to the previous year's course. In general, the students appeared to be much more active on this course than on other engineering mathematics courses, but further studies are required because of insufficient comparison data. (Majander and Rasila, 2011)

Colleagues at Aalto have undertaken a number of research projects to evaluate STACK and consider its wider place in their teaching; for example, Rasila *et al.* (2007), Rasila *et al.* (2010), Majander and Rasila (2011), Havola (2010), Havola (2011); and in the following theses: Harjula (2008), Ruokokoski (2009), and in Finnish, Majander (2010), Blåfield (2009).

7.10 Student focus group

A focus group took place with four students, two male and two female, at Aalto University in March 2012. These students were recruited as volunteers to take part in a study to better understand their experiences of learning using STACK. A semi-structured interview was chosen because of the freedom this gives to follow up themes or concerns in a way a paper-based questionnaire does not. In any case, local research such as Havola (2011) had already used questionnaires to investigate students' attitudes and learning styles. Before the interview the following questions were identified to structure the discussion. The first question is purely factual, but was included to set the scene and to settle students into the interview.

1. In which courses do you take online STACK questions?
2. In what ways do these questions help you understand what is being taught?
3. What do you most like about STACK? What do you like least?
4. What would you most like to change, and what difficulties have you had?
5. Do you use online quizzes for 'high-stakes' assessments, e.g. examinations? What concerns do you/would you have about such use?

The interview was conduced in English and audio-recorded. The audio was transcribed verbatim, and then very minimal editing undertaken to correct the English where this did not change the meaning. The complete transcript was then returned to the four participants, numbered 1–4 in the excerpts quoted below, who were given an opportunity to comment on the views they expressed or to expand on themes since the interview. A thematic analysis was conducted, and the following themes were identified. The interviewer (author) is identified by CJS below.

Feedback and flexibility

The immediate feedback was the most popular feature.

> 4: I found the immediate feedback from the STACK system very good because you can do it at home or via your cell and you know if you get it right or get it wrong and you have a number of tries. I think that was probably the best, best feature.
>
> CJS: So how does that differ from looking up the answer in the back of the text book? You could use a text book?
>
> 3: Yes, but then you look at the answer before you have solved the problem. STACK won't tell you!
>
> 4: It is a bit of cheating. You don't learn if you just go ahead and look at the back. And usually when we have homework during the course from the book they are usually problems you don't have the answers to, so you can't find out if you are wrong or right.

Other students agreed with this:

> 1: Yes, I agree with that.
>
> 2: Yes, I think it is good. Because of the feedback of course then if with the paper you have to wait and then when you see the right answers you can look through those with the teacher probably too quickly and you can't take your time to understand, but with STACK you can take your own time with those exercises. So that is the good thing with them.

Related to the immediacy of the feedback was the flexibility of access. Students all reported that they valued being able to undertake the work *'whereever you want, and whenever you want'*. Immediacy of feedback was only found to be beneficial when the feedback was specific to the task, as suggested in Section 3.6.

> 1: Sometimes it is really difficult because, ergh, it says only right or wrong. So you don't know if you have the right equation, but you have just the wrong numbers in there or it is difficult where it is wrong in the exercise, where you have done that mistake.

This kind of feedback was felt to be too strict, especially when it came to dealing with floating point numbers.

> 3: Of course, you only have one correct answer. If you round it incorrectly and the teacher does not take that into account, of course, that is a mistake even though the whole problem is solved correctly except for one rounding.

This is surely an unintended consequence. It reinforces our earlier comments in Chapter 6 that the teacher has to be very specific indeed in articulating the properties which are acceptable. Arguably two tests are needed here to differentiate between 'correct' and 'on the right track'. The author has plenty of his own experience authoring questions with such unintended *hard edges*. Numerical precision may be an underlying convention or may be explicitly stated in the question. The form of the answer might be an explicit goal of the question. Unfortunately in this case students were not happy. This is a good example of how the development of 'quality assured' questions can only be done with student involvement and the question author's willingness to engage with that involvement during subsequent academic cycles.

Students also discussed working in groups. One worked alone, while others worked in groups with friends. Interestingly, students felt able to collaborate more in groups, but each student still had their own problem on which to work.

> 4: I think one of the best things about STACK was the way it created the values, or the problems, meant for you. But they are still the same as your friend has so you can collaborate on them and do some team work, and work on the difficulty with your friends, but you still have to do the exercise for yourself [3: yeah!] you have values and
>
> 3: . . . so you can't just copy!
>
> 4: It won't help if you just copy the answer from your friend.

This has been cited by students elsewhere as a benefit. With fixed paper-based work students often worry about 'copying', and the more serious charge of plagiarism in the context of summative assessments. This is an unexpected outcome of CAA, and harnessing such collaborations in productive ways is one opportunity not available with a single paper exercise sheet, or published textbook.

Difficulties with syntax

It is no surprise that these students recalled *initial* difficulties with the input syntax. All groups of CAA users report such problems when first making use, particularly if the first use is in a high-stakes situation.

> 3: I agree, but when you get used to STACK all that goes away [4: yes] but when you start that is a problem. It is very annoying when you try to type something . . . well you have the 'check syntax', but if the check syntax is always incorrect you are like !!!
>
> 4: I think the input, the way you type in the answer. It, as we already said, it is a bit complicated. Especially if you have just started and you are used to writing things down on paper. It can be very complicated to work with all those brackets and square roots or, . . . written . . . all that. It can be frustrating at the start.

It was clear that it was syntax which was a barrier here. For one student, prior knowledge removed this problem.

> 4: Yes, I took this test but because I have a little bit of background of computer programming so I [. . .] knew the syntax a bit. I was more frustrated because I didn't know the answers myself! [2: laughs] So I guess I have time to deal with the real mathematical problems, so I guess my frustration is based on my own lack of mathematical knowledge. So, I think the test worked quite well for me. But there you have it, I had some background with things like this.

Use in examinations

Students showed a surprising willingness to participate in high-stakes automated examinations.

> 3: I would love it. [. . .] yes, a timed exam, or even an untimed exam in the way that you would have the time you want to spend with it, say you would have a day. It doesn't really matter but you would know instantly what you did . . .

Notice here that there are immediately a number of issues to unpick. Not only has the format of the examination been changed to make use of STACK, but also the length of the examination and the availability of feedback. It was clear that the students expected immediate feedback at the end of the examination.

Students raised no concerns about *validity* or *reliability* of automated examinations. When questioned explicitly about this

> 3: [. . .] I think it's great that we use machines for the things we can. I don't need a person to evaluate my skills if the computer does that as well as a person will do.
>
> And since STACK does record the answers you can have the person to go through the answers once more to check that, ok this person answered like this and the machine gave them half the points. [. . .] For the rest of the people who feel the computer is not trustable then they can have the person check out the answers. Again, I don't need that.

However, students were concerned that *method marks* would not be available, and unhappy at the lack of partial credit for correct method.

> 1: Yes, it can be ok, but I don't know if I have a problem. I'm not sure I can do it right. I get the wrong solution [. . .], but I know how to start to do it. Then I don't get half points. I have to have the right answer to get all the points. When you have a human correcting it then he or she says 'ok, this student knows how to do this but it goes wrong somewhere'.

Of course, this presupposes that there is a unique method with a set number of steps. It is also likely that in summative settings students are required to know the method as well. Students were asked if, during the examination, the system said 'no, your answer is wrong', and provided an opportunity to try again if that would address their concerns.

> 1: Erm. This could be alright, if you get feedback during the test or exam, because when you do this on paper no one is coming to say 'yes' is this totally wrong or is this the right answer. You have to think for yourself, ok I will do this now. But, yes. . . . That would be ok .. to get feedback during . . . I don't mind that.

Another student supported this idea, and went further, changing the emphasis from summative examination to a learning experience.

> 4: I think it would be very good, if . . . if you could get STACK to give you immediate feedback and make the whole exam situation like a learning experience.
>
> 3: I think it should be used a lot more, like his (4) idea that we should have immediate feedback on the exam for example. That the exam should be for learning purposes not just . . . [. . .] The exam is like you are exiting everything out of your head onto the paper and that's it. [. . .] Immediate feedback would be really good.

The students had more radical ideas, e.g. including group examinations.

> 4. [. . .] in some of our courses we have an exam you can take in a group of people. And you can evaluate what other people think and you submit one answer, and then one grade goes to the whole group, or something like that. [. . .] STACK, could give you some hint, 'you are going wrong over there', or something like this. So you could learn while you are doing the test. [. . .] If that would be possible you could get some immediate feedback and learn during the whole thing, when you are giving your answer.

Immediately a second student supported this idea.

> 2: Yes, I agree [. . .], after you have done the test [. . .] and you have got the points then you should see the right answers because when I have done exams then I'm always thinking, 'how should I have done that'. But you have to wait until next time you have the lesson, or you see the teacher, then you ask. But then you have a problem remembering what you want to know. So you see right away after you have done the exam how you should have done that problem. That would be really good.

This deserves further investigation.

7.11 Conclusion

This chapter has described the goals and design decisions which have been taken in the implementation of the data structures and interaction model behind STACK. It also reports the significant use of STACK made by Aalto in a variety of courses since 2006. STACK is perceived by staff and students to play an important role in methods-based basic teaching courses. Scores in STACK assessments correlate well with other forms of assessment. Care is needed to ensure that syntax does not constitute a barrier to students early in their studies. One finding which was surprising to the author was the acceptability of automated examinations to the members of the focus group, and this deserves further, more detailed, study. Many of these findings correspond closely to those at the author's own institution which has used CAA since 2000. In Chapter 8 we examine further examples of CAA systems and their use.

8

Software case studies

> We are still doing considerable hand filing of punched cards at this stage. This large deck of cards which includes the grader program is then run on the computer. Our largest single run has been 106 student programs covering nine different exercises. [...] We have probably checked so far some three thousand student programs using graders. (Hollingsworth, 1960)

In this chapter we examine a range of computer aided assessment systems as case studies. Starting with early history of CAA, we look at the various goals colleagues have set for themselves and the progress made. We examine the contemporary state of the art. Examples of systems have been selected to illustrate many of the common issues which others have tried to address, or because of some key novelty. In particular we examine how people have chosen to assess steps in working, or reveal sub-steps as hints. We consider some examples of expert systems and adaptive testing, and the ambitious attempts to assess mathematical proof. Needless to say, such software changes very rapidly. There are many examples in current use, and it would be both impossible and repetitive to represent them all. Hence, this chapter does not aim to be a complete catalogue.

8.1 Some early history

Computer aided assessment has a history going back over half a century. Hollingsworth (1960), quoted at the start of this chapter, was an early pioneer. He used computers to test the behaviour of students' machine-language submissions in a computing class using a 'grader' programme which automatically checked some aspects of their coding. This short paper contains many of the themes which subsequently surface regularly.

> We could not accommodate such numbers without the use of the grader. Even though the grader makes the teaching of programming to large numbers of students possible and economically feasible, a most serious question remains, how well did the students learn? After fifteen months, our experience leads us to believe that students learn programming not only as well but probably better than they did under the method we did use—laboratory groups of four of five students. (Hollingsworth, 1960)

By the mid-1960s computers were being used to teach arithmetic, such as the early work of Suppes (1967). By the 1980s the field was becoming large enough to have different branches. One came under the title of *intelligent tutoring systems* and another as *computer aided instruction*, and these drew on a tradition from the artificial intelligence community; see Sleeman and Brown (1982) and Quigley (1988). These researchers had ambitious goals, which even by the 1980s was being reassessed.

> Early CAI [Computer Aided Instruction] workers set themselves the task of producing teaching systems which could adapt to the needs of individual students. It is now generally agreed that this is a *very* difficult task, and one that will only be accomplished as a result of extensive research in both AI and Cognitive Science. (Sleeman and Brown, 1982, Preface)

Furthermore, the systems developed were often designed to tutor in only one confined area of skill-based knowledge, e.g. arithmetic, or solving quadratic equations, or symbolic integration. Examples of all these domains are given by Sleeman and Brown (1982), and they remain popular. However, by the early 1980s there was also a backlash from some about the use and, indeed as they saw it, *abuse*, of computers for testing in this way; e.g.:

> In most contemporary educational situations where children come into contact with computers the computer is used to put children through their paces, to provide exercises of an appropriate level of difficulty, to provide feedback, and to dispense information. The computer is programming the child. (Papert, 1980)

This dissatisfaction led directly to the LOGO programming language,[1] which was specifically designed to give the child the opportunity to programme the computer. The potential for using computers in this way was acknowledged much earlier by the editor's comment following Hollingsworth (1960).

> While there is a great deal of research being accomplished on 'teaching machines', many computer educators have not realized that when teaching the use of the computer they have access to the finest 'teaching machine' of all—the digital computer. This is the first of what the Editor hopes will be a fine series of papers on aspects of the education of people to use computers. – A. J. P. (Hollingsworth, 1960, Editor's note)

Programming is something of contemporary concern, e.g. Furber (2012), which has resulted in initiatives such as the Raspberry Pi foundation, and one laptop per child. LOGO, and exploratory computer environments in general, are an important part of the story, but much has been written about it elsewhere. See, for example, Noss and Hoyles (1996). Our concern here is with platforms which can be used for more formal assessments, ideally

---

1. The author was significantly and positively influenced by his experiences of working and playing with LOGO on a BBC *Model B* computer, with 32kb of memory and a cassette-tape drive for permanent storage, while at primary school in the mid-1980s.

over a wide range of mathematical topics. Another concern is the ability of individual teachers to tailor assessments to their students, and to the rather less glamorous subject of *practice*. These issues, particularly the latter, are sometimes ignored in the literature elsewhere.

This tension between learning routine techniques and exploring ideas and concepts in a less structured environment lies at the heart of mathematics education. The role of the teacher, and its complexity, have been acknowledged by many. For example:

> In the last five years researchers have focused on supportive learning environments intended to facilitate *learning-by-doing*: transforming factual knowledge into experiential knowledge. These systems attempt to combine the problem-solving experiences and motivation of 'discovery' learning with the effective guidance of tutorial interactions. These two objectives are often in conflict since, to tutor well, the system must contain the student's instructional paths and exercises to those whose answers and likely mistakes can be completely specified ahead of time. [. . .] In order to orchestrate these reasoning capabilities it must also have explicit control or tutorial strategies specifying *when* to interrupt a student's problem-solving activity, *what* to say, and *how* best to say it; all in order to provide the student with instructionally effective advice. (Sleeman and Brown, 1982, Preface)

The difference between those who focus on *skills and their acquisition* and *conceptual understanding* is a classic dichotomy in mathematics education.

8.2 CALM

The CALM CAA project started in the Department of Mathematics at Heriot-Watt University in Edinburgh in 1985, funded as part of the Computers in Teaching Initiative. Further funding, and a number of awards, enabled subsequent rounds of improvements which developed first into the CUE system in collaboration with UCLES (University of Cambridge Local Examination Syndicate) and EQL (a software company based in Livingston, West Lothian). Initially, CALM presented its tests on a PC and Mac, but from the late 1990s questions were being delivered over the Internet. Following active educational research in the early part of this century the e-assessment engine took on one further incarnation becoming known as Pass-IT, which is described in more detail in Section 8.3. In addition, a derivative 'e-assessment' engine emerged from EQL under the commercial name of i-assess which has had extensive use by professional accountancy bodies. Hence, there has been a continuous process of development and use spanning more than a quarter of a century.

The first phase of CALM replaced conventional tutorials in first-year calculus. It is interesting to recall how few tools were available to help the CAA system designer during the mid-1980s, and the limited power and sophistication of machines available to students. Even a library to *evaluate an expression at a point*, crucial to establishing algebraic equivalence in the absence of CAS, had to be written from scratch. All the major issues we address in this book had to be resolved, including input and display of mathematics, plotting graphs and diagrams, and even a 'menu' system to enable navigation around the application, see Beevers et al. (1991).

CALM was a learning and formative assessment package employing diagnostic, self tests, and monitoring tests throughout the first ten years of its use. Support for self-testing was considered by the designers to be particularly important, as it helped to give the students confidence with the system. In this mode of delivery no record of student performance in a test was kept. When the students felt more confident they were encouraged to try the monitoring tests which did record their student achievement making results available to both student and teacher.

From the outset, a serious attempt was made to automate the assessment of steps in students' working. An example, from Beevers et al. (1991), of how this worked is shown in Figure 8.1. In this mode, known as 'HARD', the student must enter answers to all parts before any feedback is given. Other modes gave feedback in different ways, and in some modes of use students could select the level of feedback they wished to receive and the level of difficulty they wanted to accept. Multiple tries, and attempts at different versions, were also possible.

> The number of attempts allowed can be varied in the software, but in most cases we set this to three and it seems to work well. (Beevers et al., 1991, p. 97)

Intriguingly this is precisely the same conclusion which the Open University used in their OpenMark system, described in Section 8.4, with progressive feedback at each stage.

As we can see from Figure 8.1, CALM is a very early example of a CAA system which moved away from multiple-choice questions and accepted mathematical expressions as answers. Inevitably, the designers had to address the issues of syntax which we discussed in Chapter 5.

QUESTION: 1

Find the derivative of the function F defined by

$$F(x) = \frac{f(x)}{g(x)} = \frac{x^2-1}{sin(x)}$$

You will now be asked to input your answer(s). 5 part(s)

```
Function in the numerator f(x) = ?   x*x - 1
Derivative of f, df/dx = ?   2x
Function in the denominator g(x) = ?   sin(x)
Derivative of g, dg/dx = ?   cos(x)
Derivative of F, dF/dx = ?   (2x sin(x) - (x^2 - 1)cos(x))/sin^2(x)
```

Are you happy with your answer(s)?   (y/n)

Figure 8.1: First phase of the CALM CAA system. From Beevers et al. (1991), Figure 4.5.

> We would like to make input simpler but have also recognized that restrictions can be advantageous. Most students will be using computers in other areas of their work and will need to learn to adapt to the rigours of computer input. The student is also forced to think much more carefully about the format of their answer, since any ambiguity is punished mercilessly. This may be frustrating at first but can lead to a better understanding, whereas a written answer may contain an ambiguity which is not recognized by the student and can lead to a misunderstanding later. (Beevers *et al.*, 1991, p. 113)

As we have seen, a typed linear syntax is one approach to the entry of a mathematical expression. Another is to provide students with a static input tool, i.e. when a mathematical expression had been typed in then a click of the mouse allowed the students to see how it was being interpreted by the computer. This input tool displayed fractions and powers in the usual way. However, during the Mathwise project, see Section 8.8, a dynamic input tool was created by the CALM team. This input tool mirrored in another window how the computer was interpreting the mathematical expression as typed keystroke by keystroke.

As early as 1991 CALM was able to do the following:

- Random generation of questions, and the random selection of questions. Random integers were inserted into an expression, but without computer algebra support to subsequently manipulate expressions this was limiting in many cases.
- Students' answers were mathematical expressions, and the system sought to establish equivalence with a correct answer (by substitution of values to variables) and the form of the expression.
- Type, timing, and the level of feedback detail could be varied.
- Questions were broken down into steps.

We note that in the early versions of CALM, each question was written as bespoke Pascal code. This required detailed technical knowledge of the system, and of computer programming. Subsequent versions of CALM enabled authoring of questions at a higher level, and with the invention of the internet the use of a Pascal application has given way to an online system.

Dealing with steps in working is perhaps the most distinctive feature of the CALM project, and subsequent child-projects such as CUE and Pass-IT. This was recognized early as a key potential educational advantage of the computer environment providing scaffolding and structure to students' answers. Of course, automatic marking also gives fast feedback to right and wrong answers and from the start students embraced this novel approach to learning.

> By giving the correct answer at each stage the student is encouraged to carry on but still has to work at each part before the answer is revealed. In a traditional text book example either the only answer given will be the final one or, if the question is answered in stages, it is difficult to avoid seeing all the answers at once. (Beevers *et al.*, 1991, p. 112)

In 1995, again as part of the Mathwise Project, the CALM team undertook an educational experiment to use the computer to grade, in part, the student performance at the end of a term of calculus with first-year undergraduates. The pilot test was designed to cover the so-called lower-order skills of knowledge, comprehension, and simple applications, and left the higher-order skills to be tested on paper in a conventional way; see Beevers *et al.* (1995) for more details. The experiment proved successful and a computer test remained as part of the measurement of student ability in all three terms for students of engineering and science at Heriot-Watt University as they studied their Service Calculus course.

CALM continues to be influential within the UK. As the developers of the recent NUMBAS system explain,

> [...] we felt that the well-used and proven CALM-style design made the optimal compromise between difficulty of authoring and robustness of design. Since no CALM-style systems matching the above criteria were available to us, we decided to start from scratch on our own system. (Foster *et al.*, 2012)

8.3 Pass-IT

The group at Heriot-Watt University has evolved since the early work of CALM to undertake a number of separate projects, including the use of CAA of mathematics in large high-stakes situations. CAA in mathematics at Heriot-Watt now takes place as part of SCHOLAR, (scholar.hw.ac.uk), a large international online learning programme. Describing themselves as a 'virtual college', they provide interactive and distance learning materials in a wide range of subject areas with a focus on STEM, business and languages. Levels range from schools to colleges and through to university.

SCHOLAR provides learning materials and services, such as CAA, in a single environment. This includes *'learning materials, activities, assessments, and revision packs'*. The platform enables monitoring of student activity and information to be provided to the teacher with tutor notes and curriculum and planning information. SCHOLAR currently has over 100,000 enrolments annually in Scottish secondary schools. Pass-IT is the assessment system used by SCHOLAR for mathematics. See www.pass-it.org.uk for more details.

Features of Pass-IT for mathematical CAA

Pass-IT, as a direct development of CALM, maintains many of the features pioneered there—in particular, randomly generated questions, answers which are mathematical expressions, and crucially, steps in the working. Figure 8.2 shows a question from Pass-IT asking the student to find the equation of a straight line in a particular form. The student could simply provide their final answer, if they do this correctly then the system will award three marks.

If they are unable to do this and ask for steps, then the steps in Figure 8.3 are shown. Now, the student is able to separately indicate the gradient and intercept. By breaking down

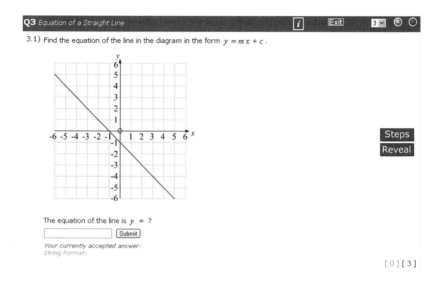

Figure 8.2: A Pass-IT question, before steps are revealed to the student.

a question in this way, partial credit is possible, and the student is reminded of how to start the question. Once this part is done, then the student can return to answer the full question. In Pass-IT, use of steps is a student choice. They are never automatically triggered, and are always optional. However, sometimes steps are chosen at the cost of some marks. Such a

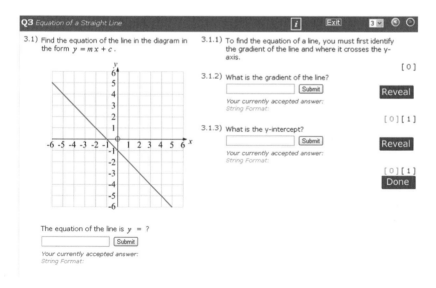

Figure 8.3: A Pass-IT question, after steps have been revealed to the student.

protocol can be used in both formative and summative mode. It is really about whether the steps are giving information for which students would otherwise have gained credit in the marking scheme. Further examples of questions with steps are given by Ashton *et al.* (2006). We shall comment on the comparability of CAA and paper-based questions when we discuss high-stakes examinations below.

Other protocols can be envisaged and once 'steps' are available at a technical level there are a number of options which a teacher might choose to use. May students request steps? If so, would they relinquish the method marks? Would steps automatically be shown if a student repeatedly makes a mistake? These options are independent of each other. STEPS could be nested, though it is likely that any student who needs more than two or three levels of steps would probably be better off attempting more elementary questions. This might include adaptive testing.

A more recent development in Pass-IT is the ability of the system to accept answers in the form of the configuration of a graph. For example, in Intermediate Mathematics 2, Unit 2, students solve simultaneous linear equations in two variables. They are expected to learn two different methods for solving such systems: (a) to solve *graphically*; and (b) to solve *algebraically*. Clearly, with this goal we need to capture both the method and the final answer.

The approach to assessing the graphical method in Pass-IT is to enable the student to plot two points on screen, as shown in Figure 8.4. The students interact with the graph by clicking points on screen to create them. Once two points have been added, the system creates a line. The student is then free to drag the points to move the line to the position they think is correct. Next, the student needs to read the coordinates of the point of intersection and enter this as part of their answer. In this way, both partial credit and follow-through marking are implemented by this multi-part question. Note that in the example shown in Figure 8.4 the student has the answer only partially correct.

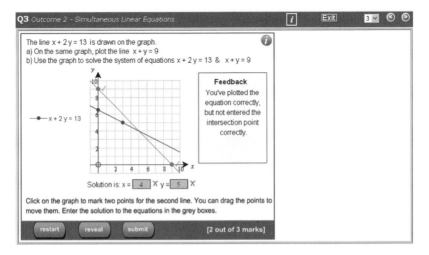

Figure 8.4: A Pass-IT question in which part of the answer is a graph.

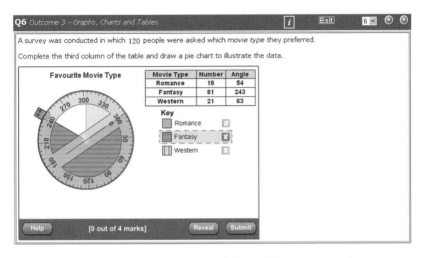

Figure 8.5: A SCHOLAR question in which part of the answer is a pie-chart.

Ashton and Khaled (2009) provide more examples where a graphical configuration constitutes the answer. These include estimating a line of best fit for a given dataset, drawing box-plots, and pie-charts. An example is shown in Figure 8.5. In developing such a graphical interface there is a balance to be struck between simplicity and structure, without giving away the method to be used. For example, to create a pie-chart, the student must complete the *angle* column of the data-table on screen. They may then drag an angle tool to create a sector. The colour of a sector enables students to specify which data entry it represents.

High-stakes examinations

One particularly interesting aspect of the work of Pass-IT is the development and implementation of large, high-stakes examinations in secondary schools.

Since 2000 the Scottish school system has used a two-stage assessment process. Typically three units make up one annual course, taking place over three teaching terms. Each unit has a final minimum competency assessment with pass/fail as the only outcome. Students must pass all three units to be entitled to take the end of course assessment. It is this end-of-course assessment which seeks to assess both lower and higher order skills. Originally, the minimum competency assessments were administered in class by the teacher with questions taken from the Scottish *National Assessment Bank (NAB)* of questions. The teacher then marks and reports to the Scottish Qualifications Authority (SQA) which of the pupils are eligible to take the end of course assessment. The SQA moderates this process. The Pass-IT project sought to replace the teacher administered and marked minimum competency assessments with CAA, the so-called e-NABs. The e-NABs consist of both formative and summative assessments for units 1 to 3 of SQA Higher Mathematics. The formative assessments are seen as vital preparation for students, and as providing valuable learning resources. This is an example where development of CAA was followed by research, which validated the equivalence of CAA with the existing paper-based tests; see Ashton et al. (2004, 2006). This developed into large-scale pilot implementation on a national scale, and in 2009-10 the Higher Mathematics e-NABs won the e-assessment Scotland awards.

In such high-stakes assessments the issue is not only formative assessment, but the need to award partial credit.

> Assessment on computer normally marks an incorrect answer wrong and awards no marks. This can lead to discrepancies between marks awarded for the same examination given in the two different media. (Ashton et al., 2006)

One approach is to break a question down into 'steps', providing one answer box for the answer to each step. As a specific example, a student might be asked (as in Ashton et al. (2006)) the following:

▼ **Example question 16**

Find the equation of the tangent to the curve $y = x^2 + 3x + 5$ at $x = 1$.

One technique for solving this question is to break the problem into the following steps:

- Find the value of y at $x = 1$.
- Find $\frac{dy}{dx}$.
- Find the gradient of the tangent at $x = 1$.
- Combine this information correctly.

It is, in some circumstances, perfectly appropriate to expect students to use a particular method as suggested here. If the goal is to obtain a correct answer and a student is struggling, then suggesting or reminding students of one method also has its place. In this case there are alternative methods: the student could simply find the remainder when the polynomial $x^2 + 3x + 5$ is divided by $(x - 1)^2$. The remainder after polynomial long division of $p(x)$ by $(x - a)^2$ always yields the tangent line at $x = a$ without using calculus; see Sangwin (2011a) for details and other methods. In most circumstances it is, with a little thought, possible to avoid unfairly penalizing a student for using a different correct method correctly.

Comparison of CAA with paper-based assessments

The group at Heriot-Watt has undertaken a number of comparisons of CAA with paper-based assessments, partly to validate their work with high-stakes examinations. For example, McGuire et al. (2002) compared the results when students took computer tests in three different formats (no steps, steps always provided or optional steps) and with the partial credit they would have obtained by taking the corresponding examinations on paper. They concluded:

There was no evidence of a difference in marks from what would be obtained from a paper-based examination or from a corresponding computer examination with steps, whether optional or compulsory. (McGuire et al., 2002)

However, 'Not surprisingly, the candidates' marks in tests without steps were much lower than those in which steps were available. They were also lower than those marks that would have been awarded in the corresponding paper-based examinations.' As Ashton et al. (2006, p. 97) subsequently commented

> [...] the authors were conscious of the fact it could be argued that not all the learning points had been examined because, in many cases, breaking the question into smaller parts gave away the strategy, which itself was a learning point.

One solution to this problem is to separate any marks for the 'strategy' and award only these marks to students who complete the question without steps. In effect, students can 'purchase' the steps at the cost of some of the marks for the question.

A separate source of partial credit in examinations is the issue of follow-through marking. That is to say, when the examiner takes a student's incorrect answer to one part and works through subsequent parts to ensure that the student's error is counted once only. This is common practice in many, if not all, high-stakes examinations in the UK school system. As we have seen, it is certainly possible with computer algebra to automate this to some extent. However, the potential for immediate feedback enables students to look through their working to try to spot such errors for themselves. This ability to try the question again could be combined with the ability to see steps, either to prompt the student with a strategy known by the teacher to be effective, or to enable the student who used this from the outset to fill in 'parts' of the question, thereby helping to isolate the source of their mistake.

Looking more widely than mathematics, Ashton et al. (2005) presented results of a comparison between paper and computer tests of ability in chemistry and computing. Their statistical model was used to analyze the experimental data from almost 200 candidates. They found that 'there is no medium effect when specific traditional paper examinations in chemistry and computing are transferred into electronic format'. They also investigated any possible effect of rewording for computer-delivered test questions, and concluded that 'no evidence of a difference could be found'.

Reporting

Reporting to the teacher is a key component of CAA, and one feature of Pass-IT is the detailed reporting mechanism; e.g. see Ashton et al. (2004). However, the *potential* for CAA to record everything has explicitly not been made available to teachers.

> It is important to allow students an opportunity to practice without being monitored in order to build up their confidence. However, equally important to student learning is feedback from their teacher. To address both needs, teacher access is provided to student results at key stages with all other results available only to the student. This means

the student can experiment and investigate without their results being available to their teacher. Then, once they feel ready, students can move on to the end-of-topic and end-of-unit assessments. The results of the end of topic and end of unit assessments are available in the SCHOLAR reporting system, allowing the teacher to track performance and progress and help to identify any misconceptions. (Ashton and Khaled, 2009)

Moving from pilots with 'enthusiast-led' assessments to mainstream reveals new challenges.

> Achievement of pedagogic validity is a necessary precursor to mainstream acceptance, but on its own it is not sufficient. It also requires well trained and informed practitioners who are able to understand and address issues relating to use of e-assessment: knowledge and understanding of assessment practice and technical skills to devise or create appropriate tests. (Ashton *et al.*, 2008)

8.4 OpenMark

OpenMark is an assessment system developed at the UK Open University, with a continuous history of related projects starting in 1974. As a distance learning institution, the Open University (OU) has a unique position in UK Higher Education. Typically, an OU course has thousands of enrolments annually. Perhaps not surprisingly, they have always had a leading role in the development of educational technology and computer aided instruction. The start of the modern-day OpenMark came in 1996/67 with the development S103 *Discovering Science*. At this time 410 questions were written for use on a PC and delivered to students on a CD-ROM. The base OpenMark library was written in C and was used by multiple other courses. Some of the basic questions are still in use in 2012, albeit in refined form. In 2000 the base library was converted to Java, allowing Internet delivery, and computation was carried out client-side, thus avoiding server-load problems. By 2002 communication of students' results to the OU server was added, and this was used for summative assessed tests.

> Open University students are studying at home, so in order to be able to move to delivering questions online (instead of on CD-ROM or DVD) we needed to be able to require our students to have Internet access, to have the confidence that students would be happy working online, and to have the confidence that our questions would be sufficiently robust when delivered across a range of systems. We had that confidence in 2002, and at that point we were able to start using online questions in high-stakes summative use on S151 Maths for Science. (Jordan, 2012, private correspondence)

Online delivery also enabled diagnostic use of questions, prior to study, with access to quizzes with titles such as '*Are you ready for S104?*'. At the time of writing this is used by around 1,000 people per month.

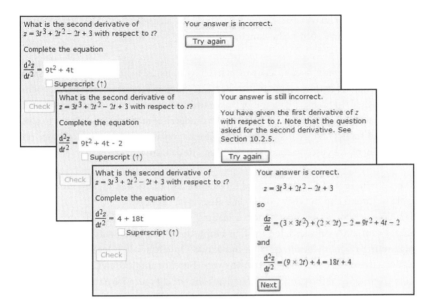

Figure 8.6: Progressive feedback from OpenMark.

OpenMark encourages multiple attempts at questions, with increasing targeted feedback. An example is provided in Figure 8.6. Usually three attempts are allowed, with the feedback after a first incorrect attempt being minimal, e.g. *'Your answer is incorrect.'* This gives students an opportunity to find and fix their own error. Even at the first stage feedback is sometimes thought to be appropriate, particularly when a student has got something wrong that they may consider to be minor, such as units. After a second incorrect attempt feedback is given that, whenever possible, is tailored to the error that has been made. When appropriate, a reference to the module materials is also provided. After a third incorrect attempt a worked example is given. It is intriguing that both the OpenMark and CALM teams independently concluded that three attempts is, in general, an optimum balance between opportunities to try again and helping students who are stuck to progress.

OpenMark makes minimal use of multiple-choice questions. In common with many CAA designers there is a strong belief that valid assessment is more likely when students construct a response for themselves rather than selecting from predefined options. In common with many systems, several variants are usually created for each question, carefully designed to be of similar difficulty; see Jordan et al. (2011). This is both an anti-plagiarism device in summative use and for extra practice in formative use.

One distinctive interface feature is the use of a very simple superscript/subscript function for students. This enables powers to be specified in a very straightforward and natural way, and this is shown on screen as the expression is typed.

The Open University has recently moved to using Moodle as an overall content management system. OpenMark is used through Moodle and is usually integrated into a module's assessment strategy with several iCMAs (interactive computer-marked assignments) as well as TMAs (tutor marked assignments). Students' attitudes to iCMAs are examined in more detail in Jordan (2011). She concludes:

> Interactive computer-marked assessment has huge potential to help distance learners to find appropriate starting points, to provide them with timely feedback and to help them to pace their study. [...] The use of e-assessment does not imply a tutor-less future, rather one in which tutors are freed from the drudgery of marking simple items to give increased support for their students, with information about student performance on e-assessment tasks used to encourage dialogue. (Jordan, 2011)

We should note that OpenMark is used for science and mathematics, and more details are available in Jordan and Butcher (2010) and Ross *et al.* (2006).

In 2006 the Open University took a leading role in developing the quiz features of Moodle. From 2009 onwards, led by Tim Hunt, the Open University undertook a major re-engineering of the Moodle quiz to introduce OpenMark-like features. This includes sophisticated response matching of text answers; Butcher and Jordan (2010). It is now also possible to implement the progressive feedback shown in Figure 8.6 natively in Moodle. For mathematics, ongoing collaboration between the Open University and the author resulted in the integration of STACK more tightly into the Moodle quiz, as described in Chapter 7.

8.5 DIAGNOSYS

Having an individual item adapt by revealing new steps is different from building up a model of students' behaviour within an expert system.

DIAGNOSYS was developed at the University of Newcastle in the UK. The last release, v3.56 in September 2007, is the culmination of more than a decade of development and use see Appleby *et al.* (1997). The DIAGNOSYS system is currently freely available, and the source of the questions can be inspected for further information. It is another example of a stand-alone application for a desktop computer. As such, it manages questions, attempts and reports the outcomes of the tests. Designed with mathematics in mind, it provides support for a range of questions, such as numeric entry or multiple choice, and also those where the answer is an algebraic expression. Hence the ever-present problems of input of expressions and the evaluation of responses need to be solved. The DIAGNOSYS system provides students with a *mathinput panel* to enable them to build up their expression in a two-dimensional format on the screen. An example expression is shown in Figure 8.7. Equivalence of expressions is established by numerical evaluation, as in many other systems. As in the CALM system, the length of the string the student types is used to establish whether it has been *simplified*. An example of a question using algebraic input, for which the form of the answer is as important as equivalence with the correct answer, is shown in Figure 8.8. DIAGNOSYS enables authors to write their own questions, and to link these questions together in quizzes.

The distinctive feature is the development of an *expert system*.

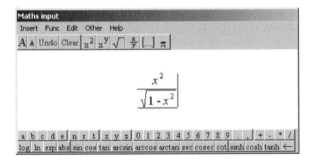

Figure 8.7: Mathinput panel from the DIAGNOSYS system.

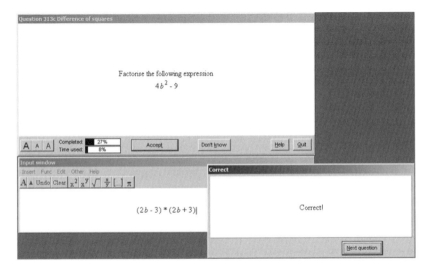

Figure 8.8: The DIAGNOSYS system.

DIAGNOSYS uses a *deterministic* expert-system. That is, inferences drawn from a student's answers are considered to be definite. Equivalently, we assume that certain skills have well-defined prerequisite skills. By careful design of both the network of skills and of the questions that test those skills, useful conclusions can be obtained [. . .]

This system is based on the assumption of the existence of a hierarchy of skills. Students start at a point determined by their last school qualification. The deterministic nature has the potential to penalise slips, or misreading of a question, in a particularly harsh manner. Hence, there is a concept of 'lives' which provides students with multiple attempts to demonstrate their proficiency at a particular skill. When a question is answered correctly then it is inferred that all linked prerequisite skills are probably known, and dependent skills (usually at a higher level) are 'possibly' known. If a question is answered incorrectly, then it is inferred that all dependent skills are probably unknown. These rules apply transitively across the network according to the partial ordering given by the arrows shown in Figure 8.9 and Figure 8.10 on the links. To select the next question the system identifies skills which are 'possibly' known, with no dependent skill which is also possibly unknown. The test terminates when there are no questions left to be asked or when a time limit is reached.

> For example, it seems sensible to assume that a student who can expand a double bracket [such as $(x + 1)(x + 2)$] can also expand a single bracket [e.g. $x(x + 1)$], whilst, conversely, a student who cannot expand a single bracket also cannot expand a double bracket.
>
> The skills were organized into a hierarchy in order that an expert system could make inferences on student knowledge from the answers previously given and select the most appropriate question to ask next. This reduces the number of questions asked for a variable ability group. (Appleby *et al.*, 1997, p. 115)

The key difficulty of this is acknowledged: '*design requires some investment of time and experience [...] Designing a good network is central to the operation of an expert-system test, which is one of the main features of the DIAGNOSYS system.*' In particular, while these maps are in terms of *skills*, in practice the system asks *questions*. While it may be educationally valid to state that one skill is a prerequisite for another it is difficult to know for certain what a particular question is actually testing. Because of the need to ensure questions are equivalent, DIAGNOSYS did *not* include randomly generated questions. As discussed earlier, randomization can easily destroy any structure, and in this case it was decided this might easily invalidate the inferences being drawn about students.

> On the whole, it seems better to keep the network moderately simple, and avoid, for example, awkward values for coefficients that are likely to give rise to wrong answers for students with a genuine, but fragile, knowledge of the skill to be tested. It is largely for this reason that random values for coefficients are not used in DIAGNOSYS; a random choice is made instead amongst variants that have all been designed carefully, and which have been validated for equal difficulty using large groups of students.

The difficulty of a question is established by looking at the success rates when used in practice in a large group. The longevity of DIAGNOSYS has made this possible. Another key point is that each question can only seek to test one skill, plus of course any prerequisite skills.

The 92 skills identified by the designers are split into the areas of arithmetic, algebra, calculus, trigonometry, statistics, and numeracy. These skills are arranged into four 'levels', which corresponds approximately to the qualifications available in UK school education. The intellectual effort needed to identify these skill maps is significant, and has been refined through many years of use with students. For the convenience of the reader we have reproduced the entire skill map as Figures 8.9 and 8.10. To give a more concrete flavour of what these 'skills' might be and how they are linked together, we consider 312, 'Expand

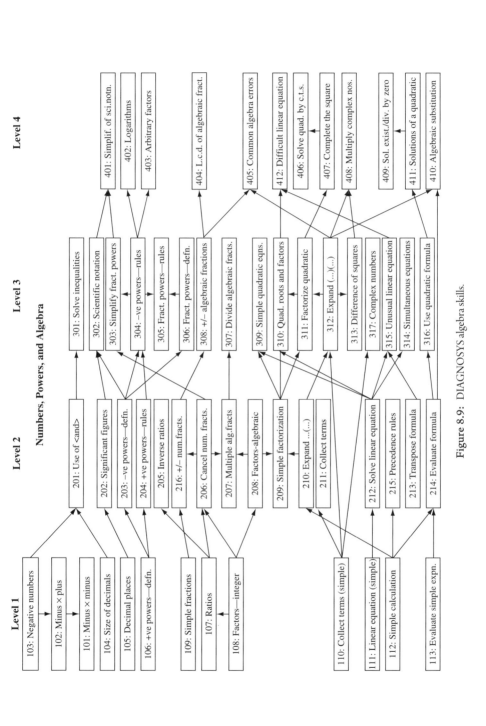

Figure 8.9: DIAGNOSYS algebra skills.

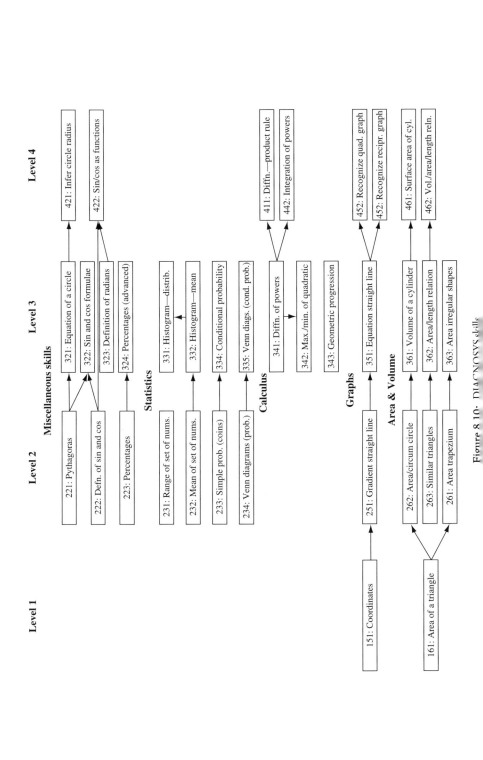

Figure 8.10: DIAGNOSYS skills

$(\cdots)(\cdots)$'. This depends on two skills: the ability to expand out single brackets as in 210 'Expand $\cdots(\cdots)$', and 110 'Collect (simple) terms'. In turn, multiplication of complex numbers (408) depends in part on this skill.

Maps such as this show explicit dependencies, which are implicit only in the normal linear order shown in Barnard (1999). Similar maps appear explicitly elsewhere, e.g. as part of the Khan Academy and in Aplusix. In capturing the network of skills, design decisions need to be made about the level of granularity. We include a further example of an expert system to illustrate this, by looking within the DIAGNOSYS skill 112: Simple calculation.

Burton (1982) identified sixty separate sub-skills within multi-digit subtraction. This includes '27. Borrow once in a problem', and '46. Subtract numbers of the same length'. The entire lattice is reproduced in Figure 8.11. The majority of these skills involve borrowing, and they are linked in a hierarchy. Assuming a student gives incorrect final answers, their

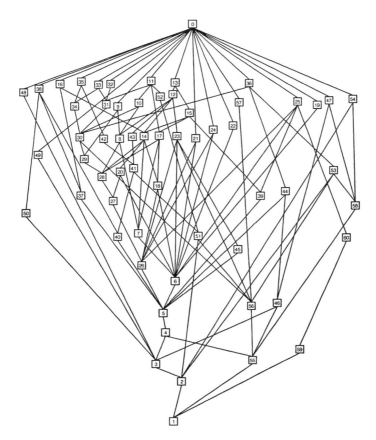

Figure 8.11: Skill lattice for subtraction. From Burton (1982). © Elsevier (1982), reproduced with permission.

goal was to identify which parts of the algorithm are not being performed correctly. Clearly this is necessary to provide specific feedback to the student. They defined a 'bug' as a 'discrete modification to the correct skills which effectively duplicate the student's behaviour', such as $0 - n = n$ when subtracting individual digits.

> [...] we were able to design a test capable of distinguishing among 1,200 compound bugs with only twelve problems! A second important property of a test is that it cause each bug to be involved often enough to determine that it is consistent. The current tests that we are using are designed to cause each primitive bug to generate at least three errors. To accomplish this it was necessary to have twenty problems on the test. (Burton, 1982, p. 172)

The validity of such tests can be confirmed, they are reliable and (subject to the availability of computing resources) practical.

The central difficulty with expert systems is designing a good network of questions. For this reason, perhaps, such adaptive testing remains a specialist area.

8.6 Cognitive tutors

The expert system approach focuses on the relationships between topics in the underlying subject matter, and pays less attention to cognition. As a response to this, from the mid-1980s onwards John Anderson and his colleagues at Carnegie Mellon University combined cognitive models of students' learning processes, see Anderson (1986), with intelligent tutoring to develop *cognitive tutors*. These are expert systems with the explicit goal of modelling students' thinking about the specific problem. It is achieved through the use of *production rules*, where each rule represents a component of the skill required to perform the task. The underlying assumption is that such rules may be mapped to cognitive skills as well as student actions.

> The core commitment at every stage of the work and in all applications is that instruction should be designed with reference to a cognitive model of the competence that the student is being asked to learn. This means that the system possesses a computational model capable of solving the problems that are given to students in the ways students are expected to solve the problems. (Anderson *et al.*, 1995)

The Cognitive Tutor uses a skills map, and students must show that they can answer questions which require skills before they can proceed to the next topic. The system selects problems for each student based on their previous performance.

> The tutors appear to work better if they present themselves to students as non-human tools to assist learning rather than as emulations of human tutors. (Anderson *et al.*, 1995)

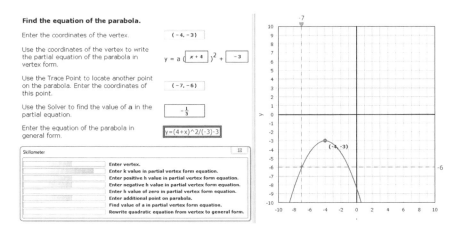

Figure 8.12: An example of the Cognitive Tutor.

Part of an example is shown in Figure 8.12. A template has been provided for steps in working at the outset, and there is an interactive diagram which the student can manipulate to find a second point with convenient coordinates to enter as another point on the quadratic. In this situation the student is also provided with access to an online CAS calculator, not shown, with which they can manipulate expressions. Notice the explicit 'skillometer', which is a pop-up window shown in the figure at the bottom left, which indicates which skills have been mastered.

The system provides hints, and as in common with much CAA, common misconceptions are coded in the background and appropriate feedback generated when a student's answer is consistent with this incorrect answer. Anderson and his colleagues at Carnegie Mellon University have undertaken a number of trials of their system. The extensive work undertaken by this group is reported online in more detail at www.carnegielearning.com. This includes psychology and the ACT-R theory of learning, evaluations of the tutors themselves, details of implementation, and more theoretical/educational papers.

8.7 Khan Academy

Many of the examples in the preceding sections are desktop applications. Over the last decade networked systems have largely replaced stand-alone applications. One of the more recent, and more popular, is the Khan Academy, www.khanacademy.org. This site provides a range of tutorial video clips, worked examples, and exercises. In particular, their 'practice' system contains a very detailed map of questions linked together as a hierarchy of skills. Many of these are mathematical, but they also include practical skills such as telling the time.

Students can ask for hints, which are considered to be an important part of Kahn Academy exercises. They are always optional and are designed to help students who are stuck. These hints are progressive, and '*The last hint of every problem should be the problem's answer.*' When students use these hints they 'set back progress' towards completing a section on the map of skills.

Exercises are marked up as HTML, and this uses client-side Javascript. Variables can be defined to enable random versions of questions. Inputs have a number of *types* which include the usual HTLM elements for multiple-choice questions. The type of the input is tied very closely to, and hence confused with, the properties expected of the answer, e.g. the *rational* type of answer compares two fractions for equality.

To manipulate mathematical expressions, the developers have begun to write libraries of mathematical functions in Javascript. These include functions such as `cleanMath(expr)` which will '*simplify formulas before display by removing, e.g., +-, --, and ^1*'. At this point little serious interest has been shown in developing CAS more fully. The majority of the questions on Kahn Academy for mathematics rely on simple types, such as numeric entry or MCQ format.

The mathematical sophistication of the Khan academy assessment system is currently best described as primitive, using virtually nothing more than string matching and insertion of random values. This is a technically simple (and robust) solution, but ignores more than half a century of research and practical development. However, what the Khan Academy does clearly indicate is (i) the demand and (ii) the popularity of such online resources. They do claim to engage large numbers of contemporary students. It also shows the level of energy and engagement which volunteers will devote to collaborative projects such as this, when the mechanisms to enable contributions are in place.

In the next section we shall return to an older project, and begin to look at how mathematical sophistication has been incorporated into various assessment systems.

8.8 Mathwise

Mathwise was a desktop computer-based learning package for mathematics, based on Toolbook, Authorware, and Hypercard. It was set up in 1992 as part of the UK Teaching and Learning Technology Programme (TLTP) and ran until December 1998. The project produced forty-eight modules covering a wide range of mathematical topics taught in first-year and second-year university courses in mathematics, science, and engineering. Modules contained material equivalent to approximately four to six traditional lectures. The modules were designed to be useful for independent learning, and so materials were closely accompanied by activities, including assessments.

Mathwise engaged forty-two authors drawn from university departments throughout the UK to develop their materials rather than have the work of development concentrated in a central team. See Harding and Quiney (1996) for more details of the project, and the list of Mathwise authors.

> One of the main aims of this initiative was to address the 'not invented here' syndrome, whereby it is recognized that CAL [computer aided learning] software is often not as easily transferable as might be hoped. (Pitcher, 1997, p. 709)

The community involvement was seen as important in the project evaluation report.

> Involving authors from a large number of universities in the construction of Mathwise materials can also be seen to have the advantage of encouraging ownership of these materials by the academic community; and it is possible that this development strategy may have been one of the factors which has contributed to the dissemination and fairly widespread use of Mathwise modules.

Furthermore

> The Consortium also favoured the *author as implementor* approach, so that at the end of the project, the necessary courseware writing skills are embedded in the community. [...] We regard the achievement of this level of collaboration, and the dissemination of courseware writing skills, as of great importance comparable to the production of the courseware itself. (Harding and Quiney, 1996)

The module on complex numbers is described by Pitcher (1997). He describes the design and also the results of a pilot study of use.

> A significant comment made by many students was that there can be a strong temptation to pass lightly over from one screen to the next and that the user should avoid doing this. [...] Arguably students should be able to use CAL software in a totally independent, self-motivated way, but the reality is that many students lack the capability to do so, at least on entry to higher education. This is an issue of study skills. (Pitcher, 1997, p. 714)

> A somewhat disquieting tentative conclusion, relevant to this cohort of students at least, is that the utilization of a learning resource seems to be inversely proportional to the amount of initiative required to use it. (Pitcher, 1997, p. 715)

In looking for quantitative evidence for an effect he compared those students who used Mathwise with those who did not, acknowledging the potential for the 'Hawthorn effect'.

> The twenty-six students who have used Mathwise scored an average of 64.0% in the whole examination, compared with an average of 58.4% among those who had not used Mathwise. This represents an improved performance by a factor of 1.096. [...] The same improvement is seen, both in the specific topic of complex numbers, and also in the mathematics paper as a whole. [...] It is therefore difficult to claim that Mathwise in itself is inherently superior to other learning resources. What can be argued is that, as an alternative resource, it can be equally effective [...].

The assessment component of Mathwise was developed, in part, by the team at Heriot-Watt, based on their experience with CALM; see Harding and Quiney (1996) and Beevers *et al.* (1995). Further technical innovations were made by Michael McCabe of the University of Portsmouth, which incorporated CAS technology for the first time.

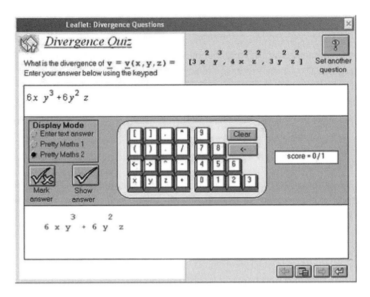

Figure 8.13: Mathwise vectors module: the divergence quiz.

Back in 1995 I was offered the Maple kernel to hook into Toolbook! It was called 'Mathedge' and it gave me the ability to call Maple V3 commands from within the authoring software. I slaved long and hard to overcome technical problems, but eventually it all worked perfectly. I had the ability to create a slick user interface (à la Mathwise) with mathematical power behind it. I began developing all sorts of clever stuff from what I called *algebrators* to CAA, using Maple as the assessment engine. It had been previously possible to write Maple programs to do simple 'e-assessment' but this was the first time anyone had created CAS powered e-assessment as we now know it. (McCabe 2010, private correspondence)

An example is shown in Figure 8.13. In the next sections we see how CAS continued to be used for CAA in this way.

8.9 WeBWorK

WeBWorK is an online CAA system targeted at mathematics and science in undergraduate courses, Gage *et al.* (2002). Initiated by Arnold Pizer and Michael Gage at the University of Rochester in 1995 as a development of the CAPA system, it has been subsequently developed and maintained by the mathematics community as an open source project. WeBWorK was built on freely available web technology, and the software is claimed to be used by more than 240 colleges and universities. Combining technologies in this way, rather than writing dedicated desktop software, was rather innovative at the time. The modular construction and extensibility, both of the underlying mathematical software and front end, have enabled WeBWorK to evolve more or less continuously for the last fifteen years. The American Mathematical Society survey of the use of homework systems found that WeBWork was the most popular. '*Among faculty-created software, WeBWorK, developed by University of Rochester mathematics faculty, has been the most successful*', Kehoe (2010). She found that this software was cited 55 times as being in use by responders to this survey. '*Top-eighty doctoral departments were twice as likely to use WeBWorK (almost 40%) as M.S. or B.S. institutions*', and Kehoe (2010) found that more than 100,000 students have used WeBWork. Hence, this project has both a long track record, and a large community of practice, many of whom are in the USA. WeBWorK is supported by the Mathematical Association of America (MAA) and the US National Science Foundation. Currently, the MAA offer limited hosting of WeBWorK courses for departments unable to support a server of their own.

Students must enter their answer using a typed linear syntax into a web page, much as in STACK and many other systems. WeBWorK also implements a *preview* protocol to enable students to view their expression before it is assessed. Some interesting design decisions have been made with the input syntax choices. For example, multiplication can be indicated by juxtaposition: 2x, explicit multiplication using a star: 2*x, or the use of a space character. For example, 3 4 is three multiplied by four, not 34. WeBWorK supports interval notation in the following form:

```
(-Inf, 3)U(3,4]
```

This denotes the union of the open real interval $(-\infty, 3)$ with the half closed interval $(3, 4]$. Also there is strong support for scientific units. A detailed analysis of all student entries in problems that required formulaic answers is given by Roth *et al.* (2008), who also undertook a survey to examine student perceptions of WeBWorK.

> We found that the immediate feedback provided by the system and ease of access were strongly appreciated, while the majority of the complaints were related to syntactic difficulties. (Roth *et al.*, 2008)

They go on to conclude as follows:

> Traditional methods of analysis have neglected an important source of data in web-based programs of this type: the complete record of student entries collected in online sessions. Since students often submit multiple attempts for each problem, these records can be analyzed in the context of student problem-solving, revealing subtle aspects of student problem-solving behaviour that are inaccessible by traditional study methods. We argue that the analysis of student entry errors, although more nuanced and time-consuming than examining scores, grades, and student opinion, is valuable for understanding both learner strategies and any possible obstacles unintentionally introduced by the system. (Roth *et al.*, 2008)

This corresponds exactly with the author's experiences with the STACK system, and is a common theme emerging from the disparate CAA systems currently in use.

Dedic *et al.* (2007) contrast students' performance and persistence in (C1) traditional lectures with paper-based assignments, (C2) traditional lectures with WeBWorK assignments, and (C3) traditional lectures with in-class interactive sessions, designed to support students working on WeBWorK assignments. Their study involved 354 students and eight instructors, split approximately evenly between these groups. There were no significant differences between the three groups in their prior mathematical achievement. This is an unusual example of a controlled trial in educational research looking at the effectiveness of CAA.

> C3 students significantly outperformed C1 and C2 students on all measures. They were more likely to have higher final grades (Final Grade), higher knowledge of Calculus (Final Score), higher percentage of correctly solved assigned problems (Assignment), submit assignments more frequently (Frequency) and a higher probability of enrolling in subsequent mathematics courses (Probability Perseverance). [...] Failure rates also differed significantly across the three conditions (Pearson Chi-square = .022). In condition C1, 43.2% failed, while only 36.0% of C2 students failed, and 26.2% of C3 students failed. (Dedic *et al.*, 2007, p. 6)

By controlling carefully the instructional design differences between control and experimental conditions, these researchers are cautiously confident in the robustness of their quantitative results. They found that '*there were no significant differences in performance or perseverance of students in the more traditional C1 and the WeBWorK C2*'. However, '*C3 students outperformed C1 and C2 students on every measure*' and these students were much more likely to subsequently continue with mathematical courses. The authors acknowledge the lack of a group taking traditional lectures with in-class interactive sessions, but conclude that '*these results support the conclusion of Lowe (2001) that CAI is not a panacea, but rather a tool that can enhance an effective instructional strategy.*'

> We note that virtually all instructors in this experiment were sufficiently impressed with the C3 instructional design that they now employ it in their classes. This result alone is extraordinary because recommendations flowing from educational research usually have little impact on teaching in sciences and mathematics. (Dedic *et al.*, 2007)

One distinctive feature of WeBWorK, which may be contributory to the extent of its widespread adoption, is the National Problem Library (NPL), recently renamed to Open Problem Library, of over 20,000 homework problems.

> Problems in the NPL target most lower division undergraduate math courses and some advanced courses. Supported courses include college algebra, discrete mathematics, probability and statistics, single and multivariable calculus, differential equations, linear algebra and complex analysis.

The source code of these problems is available freely to be used and modified. With such a large database of problems, actually finding a relevant task for students can be difficult. To address this, problems in the NPL are, ideally, tagged following a hierarchy of 'course', 'chapter', and 'section'. For example,

```
Calculus -> Partial Derivatives -> Lagrange Multipliers
```

This division into 13 course, 96 chapters and over 500 sections provides a fine-grained arrangement of the most common curriculum topics. The issue of how to arrange such large banks of questions arises in any substantial collaborative CAA project. This tree structure is natural, and the advantage of such a scheme is that every question belongs to a unique section. However, it lacks the richer linking structure of the Khan Academy web of skills, or the smaller dependency graph of DIAGNOSYS shown in Figure 8.9 and Figure 8.10.

WeBWorK is an open-source system written in Perl and TeX. Question authors must write in a script language called *Problem Generation* (PG), based on Perl, to create and edit questions, although many examples are available from the National Problem Library which can be used as starting points for modification. Originally written by Michael Gage and Arnie Pizer, this system has been extended by Davide Cervone with his *MathObjects* libraries. MathObjects essentially replicate CAS features in Perl. For example, mathematical expressions can be manipulated by operations such as evaluation at a point, substitution, and 'reduce', which removes items such a multiplication by one and addition of zero. Currently the library supports *types* of object which include numbers, vectors, matrices, sets, lists, intervals etc. And, as is necessary for CAA, there are a number of *contexts* which provide meaning for ambiguous symbols such as i, and set options such as the numerical tolerance for comparison of expressions. Thus, while WeBWork is not underpinned by a mainstream CAS, many of the MathObject libraries recreate computer algebra function appropriate to automatic assessment. This includes the functionality to establish the equivalence of mathematical expressions. Central are *Answer Checkers* which establish the properties of expressions and this includes both predicate and comparison functions.

> The second decision was NOT to use CAS to evaluate the answer strings—because most of the difficulties we experienced were with string handling and preprocessing of student answers, not the evaluation. The perl parser worked fine for the initial version of WeBWorK, it was later replaced by a more flexible parser in MathObjects which gives better error messages among other things. For the kinds of math used in homework Maple and Mathematica or Macsyma were overkill in terms of capability and used a lot of resources for the time. On the other hand a lot of string processing was necessary to insure that the student's answer was in the correct format and the regular expression capabilities of perl were unmatched by any CAS of the time. We still find that comparing two functions by evaluating them at 4 or 5 points is a more reliable check of equality than symbolic comparisons. We also discovered that MapleTA uses this technique often as well! (Gage 2012, private correspondence)

More information about WeBWork is available online at http://webwork.maa.org/wiki/.

8.10 MathXpert

MathXpert, and its precursor Mathpert, are tutoring systems for mathematics, designed by Michael Beeson at San Jose State University. Development began in 1985, and the current version is commercially available through www.helpwithmath.com. This is a stand-alone desktop system which also provides a strong graphical library. MathXpert allows its user to solve mathematical problems by constructing a step-by-step solution. MathXpert contains a wide range of problems with a comprehensive coverage of algebra, pre-calculus and calculus. This includes work with trigonometry and inequalities.

> *Mathpert* is intended to replace paper-and-pencil homework in algebra, trig, and calculus, retaining compatibility with the existing curriculum while at the same time supporting innovative curriculum changes; to provide easy-to-use computer graphics for classroom demonstration in those subjects, as well as for home study; to replace or supplement chalk-and-blackboard in the classroom for symbolic problems as well as graphs. (Beeson, 1998)

Students pick a topic, which enables them to choose the kind of problem they wish to solve. They then make use of the *problem selection* window to choose the precise problem and MathXpert provides a large number of sample problems of each type. Alternatively, students are able to enter whatever they wish to solve. The facility for users to specify their own problem is an unusual and distinguishing feature of MathXpert.

Having done this, students use the *calculation window* to solve the problem in a step-by-step fashion. To take a step, users select part (or all) of an expression. The system then responds with a menu of operations which can be performed on that selection. Having picked an option, MathXpert then performs that operation automatically. The selection mechanism is operated by using the mouse to draw boxes around parts of an expression. A typical calculation window is shown in Figure 8.14 in which part of an algebraic fraction has been selected. The options on the right describe manipulations which are possible for this selection.

MathXpert also has options to provide users with hints, to show the next step automatically or even to complete the whole problem. Notice that MathXpert has encoded the methodology to solve the majority of problems in elementary mathematics, including those entered by users. MathXpert enables *multiple solution paths*, indeed any correct solution can be carried out using MathXpert. While the software will only provide one solution path, by taking an alternative initial step then asking for the complete solution it is possible to generate other routes to the correct solution.

MathXpert also claims to be '*never more than one click away from a relevant graph*'. The software provides a wide range of graphical support, including implicit, parametric and polar plots. There are also relatively common, but specific, graph types such as step plots for approximate integration and the illustration of the complex roots of a polynomial. A comprehensive range of options for changing the size and appearance of the graphs is provided. The design of the graphical libraries, to ensure reliability, is described in Beeson (1998).

8.10 MATHXPERT

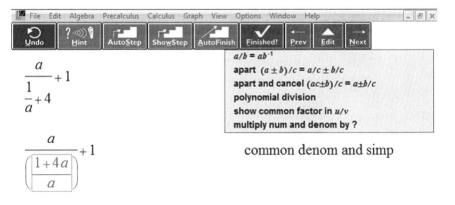

Figure 8.14: The user interface in the MathXpert system.

The distinctive feature of MathXpert is its fundamental approach to the mathematical underpinning. This goes well beyond the use of an existing CAS, such as Maxima in STACK, as a library of functions.

> ... if we start with an *educational* purpose, [...] it is impossible to achieve ideal results by tacking on some additional 'interface' features to a previously existing computational system. To put it another way: it is not possible to entirely separate 'interface' considerations from 'kernal' considerations. (Beeson, 1998)

Beeson (1998) then describes *eight fundamental principles* which guided his design. Key to this is the requirement that MathXpert should never produce an incorrect result. In particular, '*you cannot perform a mathematically incorrect operation or derive a false result . . .*'. This is the *correctness* principle. This is difficult to achieve, as is testified by Stoutemyer (1991), and for reasons we have already examined in Chapter 6. The essential difficulty is assumptions about symbols. To enable transformations such as $\sqrt{x}\sqrt{y} \leftrightarrow \sqrt{xy}$ or division by a symbolic constant which might subsequently be assigned the value zero, we have to make and track such assumptions. MathXpert keeps careful track of side conditions generated in the course of a calculation. These are available on request to a user, but are not shown by default. An auto-generated solution is shown in Figure 8.15. Notice that in removing the surd term in the first step the system has kept track of the assumption that $9x - 20 > 0$. This is subsequently used in the last step to provide a correct and complete solution. Tracking such side conditions is a non-trivial problem. E.g. when ensuring the denominator of an expression is non-zero we must deal with finding the zeros of a function. '*The correctness principle cannot be "added on" to a system that was not designed from the start to support the maintenance of a list of assumptions during a calculation*'; Beeson (1998). The correctness principle also extends to accuracy of graphs, e.g. in dealing with discontinuities and singularities.

From a pedagogic perspective, Beeson (1998) notes that it is important to solve problems in the way students would. This is what he calls *cognitive fidelity*. '*This means that it takes the same steps as the students should in a correct order*'. However, the student should

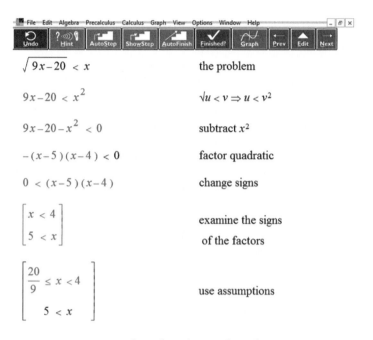

Figure 8.15: Tracking side conditions in the MathXpert system.

also be able to see how the system has solved a problem. This is the *glass box* principle: 'this turned out to be not only the most important but also the most difficult criterion to satisfy. [. . .] in practice this means accompanying each step with a "justification".' Examples of justifications can be seen in Figure 8.15. This, he claims, is the largest barrier to successful use of mathematical software by students. In particular, the traditional CAS uses a high-powered algorithm to arrive at the result of a calculation and is incapable of subsequently breaking this computation down into intelligible steps.

It is also a principle that the *user is in control* of the software and that it be *customized to the level of the user*. This is essentially a corollary of *cognitive fidelity*. 'It follows that the programme must contain some, albeit rudimentary, representations of it's users knowledge.' If the system is to be customized to the level of the user then they should be in control. '[. . .] the user decides what steps should be taken, and the computer takes them.'. Then, to be of help, the computer can take over if the user is lost. '[. . .] this is a very strong requirement in view of the fact that what is required is not just the answer, but an economical, beautiful, cognitively faithful step-by-step solution tailored to the level of the student.' Beeson (1998) fully acknowledges the theoretical difficulties here: 'we do implicitly teach methods, not just a collection of specific examples, and the requirement is simply that the programme must embody all the methods we teach'.

The last two principles are *ease of use* and that it be *useable with standard curriculum*. 'This criterion is perhaps more controversial than my other criteria. But I have always believed that serious curriculum change will be driven by students use of software, not the other way round'; Beeson (1998).

One difficulty, highlighted by Nicaud *et al.* (2004), with the menu-driven application of rules occurs when a single rule is applicable to different sub-expressions. For example, in the expression $x^4 - x^2 - 9$ we may apply the rule $a^2 - b^2 \to (a-b)(a+b)$ in two different ways, either matching to $x^4 - x^2$ or to $x^4 - 9$. In this situation it is difficult to select the sub-expression $x^4 - 9$ from $x^4 - x^2 - 9$ without re-ordering some of the terms.

We note that a substantial proportion of the MathXpert code is devoted to deriving the correct step-by-step solutions.

> *Mathpert* has several thousand lines of what amounts to *rules for using rules*. These rules are internal rather than visible to the user, and govern the machinery of the automode solution-generator. The process of developing and fine-tuning these meta-rules took much more time than the implementation of individual operations for computation, which were comparatively simple. (Beeson, 1998)

It is highly unusual to mix automatic theorem proving with computational CAS components in this way, either for the educational goals of MathXpert or indeed in any mathematical tools. Few other CAA designers (including the author) have had the courage, or expertise, to write such mathematical libraries for an educational assessment/tutoring system starting from square one. Where designers have attempted to provide mathematical sophistication this has most often been in linking existing CAS functionality. Beeson's work sets the standard in this area, and a more detailed discussion of these issues is given by Beeson (1998) and Beeson (2003), which are essential reading for anyone seriously interested in the design of CAA for mathematics.

More recent work has extended these ideas by enabling experts to encode the strategies for solving types of problems, see Heeren *et al.* (2010). Working at the strategy level enables feedback of various kinds to be triggered automatically by properties of the strategy.

> Checking whether or not a step is valid amounts to checking whether or not the sequence of steps supplied by the student is a valid prefix of a sentence of the language specified by the CFG [context-free grammar] corresponding to the strategy. As soon as we detect that a student no longer follows the strategy, we have several possibilities to react. (Heeren et al., 2010)

Hence, checking whether a student's solution corresponds to the strategy is essentially a parsing problem. Forms of feedback include progress towards the goal or using the outcomes of a partially correct answer to trigger adaptive testing. At its most basic form, the strategy can be used to trigger an automatically generated hint.

8.11 Algebra tutors: Aplusix and T-algebra

Our focus so far has been on *assessment* rather than *tutoring*. There is considerable overlap, but also a subtle but important change of emphasis between the two. Tutoring systems contain formative assessment, and assessment is wider than just formative work.

8. SOFTWARE CASE STUDIES

Aplusix is a stand alone software application for assessing arithmetic and algebra, written by Jean-François Nicaud, Denis Bouhineau and Hamid Chaachoua. Work began in September 2000 at the University of Nantes, moving in September 2001 to the Leibniz laboratory at Grenoble. It is networked to allow teachers to communicate with students. Unlike in MathXpert, where the student selects the rules and the machine performs the operations, in this 'Interactive Learning Environment' (ILE) students are expected to perform the algebraic moves themselves.

> The first main goal of the designers was to develop an ILE for algebra allowing the student to freely build and transform algebraic expressions, and providing epistemic feedback that can help in learning algebra. (Nicaud *et al.*, 2004)

A screen shot is shown in Figure 8.16. One significant feature of this system is the advanced input tool for mathematical expressions. This supports students in entering their expression, cutting and pasting between lines of calculation and editing.

Aplusix enables students to edit mathematical expressions to let the student develop his or her own reasoning steps. From this the system generates feedback which shows the state of expressions and the correctness of the student's calculations. In particular, Aplusix focuses on *Reasoning by Equivalence*.

> Reasoning by equivalence is a major reasoning mode in algebra. It consists of searching for the solution of a problem by replacing the algebraic expression of the problem by equivalent expressions until a solved form is reached. The importance of this reasoning mode comes from its capacity to solve many problems and sub-problems in algebra. However, there are other reasoning modes such as necessary conditions, sufficient conditions, recurrence, and so on. [...] There is an equivalence relationship between expressions:

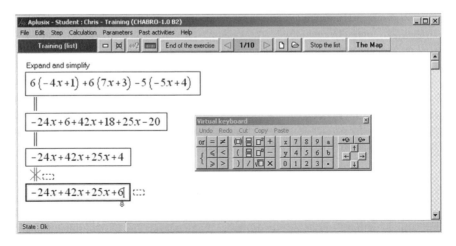

Figure 8.16: The Aplusix system.

replacing an expression or a sub-expression in a problem by an equivalent expression provides a new problem having the same solutions. (Nicaud et al., 2004)

To successfully reason in this way students must *recognize* (match) which sub-expressions can be transformed, and determine the applicable rules. In MathXpert students select sub-expressions and the system lists available rules. That is, students choose the *strategy*. Then the rule must be applied either by the student, as in Aplusix, or by the software, as in MathXpert. When reasoning by equivalence *backtracking* is the strategic decision to halt an unpromising line and go back to a previous step to try another transformation rule.

An example of backtracking is shown in Figure 8.17. Expressions are placed in boxes and lines are drawn between boxes. The boxes and lines link expressions, and the type of the line indicates if the expressions are equivalent. In the left column, the student subtracts the right-hand side from both, but then makes a mistake expanding out the brackets. This is indicated by the cross through the equivalence sign. They then backtrack to the original problem and try a different strategy, collecting terms with x on one side, and numbers on the other.

Problems are hard-coded as templates from which random versions are given to the student, currently about 400 such 'patterns of exercises' are available. Evaluative assessments are available to the teacher. There are various modes in which this software can be used, which provide different feedback regimes. For example, with *continuous verification* the equivalence of expressions is recalculated with every modification. With *verification on demand* this feedback is delayed, but it also possible to switch off all verification. Nicaud et al. (2004) also report pilot studies with students.

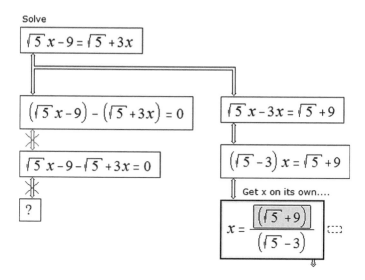

Figure 8.17: Backtracking in Aplusix.

Aplusix shares much in common with the T-algebra project, Prank (2008). In Prank (2011) the design principles are discussed including his *ten most important functionalities* for such a system which echo many of the themes addressed already, such as input, editing parts of expressions and operating on sub-expressions. Interestingly, he also observes that

> for a better indication of the location of errors, it is sometimes necessary to additionally test the equivalence between smaller components of the entered result and the corresponding parts of the previous expression. (Prank, 2011)

8.12 Conclusion

In this chapter we have described a number of CAA systems, from the very early attempts of Hollingsworth (1960) (with punch cards) through to the mathematical sophistication of MathXpert and the popularity of the contemporary online Khan Academy. There are an increasing number of commercial tutoring systems, many tied to existing textbooks, such as Aleks, Carnegie, MyMathLab, Knewton, MapleTA, and MyMath. No doubt there are many others.

Some common themes have emerged from these endeavours.

- Students appreciate the immediate feedback.
- Freeform steps in working are still difficult to automate, or tutorial systems working step by step are confined to specific topic areas.
- The syntax is regularly cited as problematic by students, and an important learning goal by staff.
- Mathematical sophistication is not tied to popularity.
- Replication and reinvention is rife.

The majority of CAA developers have sought to evaluate the effectiveness of their systems in raising students' attainment and engagement with mathematics. Methodological difficulties, e.g. lack of controlled trials, immediate or short-term studies, and developer as researcher do bring the robustness of research into question. However, clear themes can be perceived. The strongest theme is the importance of the teachers and the way they choose to use CAA. For example, in her evaluation of ALTA, McAlister said the following:

> The evidence presented in the findings suggests that the relationship between the teacher, the pupil, and the management of the system was crucial in retaining mid to high levels of engagement with the formative process. Placing the interaction between teacher, pupil, and the assessment task in the context of good teaching and learning is at the centre of the formative assessment process. (McAlister, 2005)

These conclusions are supported by Dedic *et al.* (2007), and others.

Many of the systems discussed here are the result of very long-term projects. Heriot-Watt, Carnegie Mellon, and the Open University each have undertaken continuous research and development work in this area for more than a quarter of a century. The author's own STACK system has been in development for nearly a decade. Anyone interested in creating their own CAA system is advised not to underestimate the effort required both to develop a pilot system, and to move from a pilot project to a fully-fledged robust product. A further discussion of this issue is undertaken by Oettinger (1974).

9

The future

This chapter tries to articulate some of the challenges which remain for the future, and is therefore in part, inevitably, of a speculative nature. The CAA systems in Chapter 8 predominantly seek to assess only the final answer, often using some kind of computer algebra. To assess steps in the working, fixed templates are often provided. To assess extended arguments and proof it is necessary to include some automatic theorem proving (ATP) tools. Indeed MathXpert, described in Section 8.10, did precisely this. We have already commented in Section 5.7 of the difficulties of ATP interface design and in Section 6.12 on the need to track logic within calculations. We pick up these issues in this chapter, and consider how to assess extended mathematical arguments and possibly even proof.

9.1 Encoding a complete mathematical argument

A very wide range of users need to type simple mathematical expressions at the keyboard, and link these together into the kinds of basic mathematical arguments which are ubiquitous in the early years of university mathematics and computer science courses. It is remarkable and somewhat embarrassing that in 2012 such mathematical fragments cannot be encoded routinely to be evaluated and exchanged in a standard format which is simple for humans to use. Given the pioneering role of these disciplines in creating the technological revolution of the last fifty years this is particularly surprising. In this section we provide example mathematical arguments, but there are still serious barriers to creating interactive documents to represent them. TeX and LaTeX remain the industry standards for research publications and typesetting large structured documents such as books. See, for example, Knuth (1979). The essential idea was to provide mechanisms for arranging boxes on a two-dimensional page and for filling these boxes with typographical characters. The goal was never to capture any *meaning*.

The serious theoretical difficulties in *assessing* complete arguments do not obviate the need to encode them. Many individual steps within a larger (potentially undecidable) argument *can* be checked automatically. The difference between 'no problems found' and 'no problems exist' is important, but *failure* of the former means that the proposed argument must be flawed. Some arguments, such as exhaustive cases or proof by induction have a template-like structure which could be captured, if not verified. The scope of bound variables can be encoded. All this would be very useful for a human to know. Hence, as long as users understand these issues, proof checkers are potentially very valuable additions to the CAA landscape. Automatic assessment may ultimately prove impossible, but semi-automatic assessment appears to be a very promising compromise.

We shall now provide some examples of the kinds of problems which students need to solve routinely, but for which at least two reasonable and valid solutions exist. Hence, unless the teacher requires a particular method a CAA system assessing arguments in general will need to accept them all. We will then discuss the extent to which existing CAA systems can assess the *arguments* given. All these problems form a core part of many current courses, and so would be perfectly reasonable ambitions for a mainstream general mathematics assessment system which sought to assess proof and argument.

Solving a quadratic equation

Example Problem: Solve $x^2 - 4x - 21 = 0$.

This is a classic topic in algebra, and we provide two solutions.

1. Factoring
Since $x^2 - 4x - 21 = (x+3)(x-7)$ it follows that $(x+3) = 0$ or $(x-7) = 0$. Hence $x = -3$ or $x = 7$.

2. Completing the square
$x^2 - 4x - 21 = (x-2)^2 - 25$. Hence $(x-2)^2 = 25$. Taking the square root, we have $x - 2 = \pm 5$. Thus $x = 2 \pm 5$, i.e. $x = -3$ or $x = 7$.

In Section 6.12 we discussed the difficulty of separating out pure calculation from logic, and the need to track side conditions. These example solutions illustrate these difficulties. Both solutions move beyond *reasoning by equivalence* as in the Aplusix system, see Section 8.11. The MathXpert system discussed in Section 8.10 seems to have the strongest features for encoding and dealing with arguments such as these. Many existing CAA systems, including STACK, can provide the method for students in the form of a template for this kind of problem, and this is the approach routinely taken.

Tangent lines to a curve

The next example calculus problem is as ubiquitous as solving a quadratic equation.

Example Problem: Find the equation of the tangent line to $p(x) := x^3 - 6x^2 + 10x - 3$ at $x = 2$.

1. Calculus
The tangent line has the form $y = mx + c$. To find m we differentiate, $p'(x) = 3x^2 - 12x + 10$. Evaluating $p'(2) = 12 - 24 + 10 = -2$, so $m = -2$. At $x = 2$ we have

$$y = -4 + c = p(2) = 8 - 24 + 20 - 3 = 1.$$

Hence $c = 5$. This gives the tangent line as $y = -2x + 5$.

2. Writing the cubic 'about the point $x = 2$'
Since we are interested in the point $x = 2$ we shift the origin to this point by evaluating $p(x + 2)$ and expand out the brackets.

$$p(x + 2) = (x + 2)^3 - 6(x + 2)^2 + 10(x + 2) - 3$$
$$= x^3 - 2x + 1.$$

Next we truncate this expression and notice that the best fit line at $x = 0$ is the line $l(x) = -2x + 1$. We shift this back to give

$$l(x - 2) = -2(x - 2) + 1 = -2x + 5,$$

which is the tangent line to $p(x)$ at $x = 2$.

3. Remainder on division
Divide $p(x) = x^3 - 6x^2 + 10x - 3$ by $(x - 2)^2 = x^2 - 4x + 4$.

$$\begin{array}{r}
x - 2 \\
x^2 - 4x + 4 \overline{\smash{\big)}\, x^3 - 6x^2 + 10x - 3} \\
\underline{-x^3 + 4x^2 - 4x } \\
-2x^2 + 6x - 3 \\
\underline{2x^2 - 8x + 8} \\
-2x + 5
\end{array}$$

The remainder is $-2x + 5$, which is the tangent line.

Methods 2 and 3 rely, essentially, on Taylor's theory; see Sangwin (2011a). It is most likely that students will be taught, and expected to use, the first method. As with solving a quadratic, CAA would provide a template. The alternative methods illustrate the fact that human mathematicians *compress* concepts into definitions and theory, such as Taylor's theorem. However, automatic theorem provers work from axiomatic foundations in an inflexible way. To use current automatic theorem provers it is also necessary to reformulate statements. For example, quantifier elimination is a non-trivial reformulation of mathematical statements into an equivalent form, solely for the purpose of using ATP software. The alternative option is that such processes need to be explicitly part of the undergraduate curriculum.

Integral of a trigonometrical expression

Example Problem: Calculate $\int \cos(t) \sin(t) dt$.

1. Substitution
First we let $u = \sin(t)$ then $\frac{du}{dt} = \cos(t)$ and so

$$\int \cos(t) \sin(t) dt = \int u \frac{du}{dt} dt = \int u\, du = \frac{u^2}{2} + c = \frac{1}{2} \sin^2(t) + c.$$

2. Rewrite the integrand using a trigonometric formula

$$\int \cos(t) \sin(t) dt = \int \frac{1}{2} \sin(2t) dt = \frac{-1}{4} \cos(2t) + c$$

$$= \frac{-1}{4}(1 - 2\sin^2(t)) + c = \frac{1}{2}\sin^2(t) - \frac{1}{4} + c.$$

3. By parts
Recall the formula for integration by parts $\int u'v = uv - \int uv'$. We assume $v = \sin(t)$ and $v' = \cos(t)$ so that $v = \sin(t)$ also. Let

$$I = \int \cos(t) \sin(t) dt.$$

$$I = \sin^2(t) - I.$$

Hence $I = \frac{1}{2} \sin^2(t) + c$.

The first of these solutions requires the user to specify a substitution; i.e., they need to introduce a new term and make use of this in calculations. The second method requires the *recognition* of the applicability of a particular trigonometrical formula. The last solution uses integration by parts, and the need to avoid a non-trivial circular calculation.

A locus problem

Example Problem: A ladder stands on the floor and against a wall. It slides along the floor and down the wall. If a cat is sitting in the middle of the ladder, what curve does it move along?

1. Coordinates
Imagine a coordinate system where the floor is the x-axis and the wall the y-axis. Let the coordinates of the cat be (x, y) and the length of the ladder be $2l$. Let θ be the angle of the ladder with the horizontal. Then $y = l\cos(\theta)$ and $x = l\sin(\theta)$. Squaring both equations, and using $\cos(x)^2 + \sin(x)^2 = 1$ we have $x^2 + y^2 = l^2$, so that the cat must sit on the circle. In fact, the constraints enable the cat to reach only a quarter of the circle.

2. Geometry
Assume the length of the ladder is l. Draw a segment of length l through the cat, with one end at the join of the wall and floor. It is easy to argue that this forms an X-shape, e.g. by

similarities. Hence, the cat is a constant distance from the join of the wall and floor. It moves on a quarter of the circle.

This problem contains a non-trivial modelling element. For example, the situation has not already been described in mathematical terms which are immediately amenable to mathematical computation, either with a CAS or ATP software. It also has a purely geometrical second solution.

9.2 Assessment of proof

As discussed in Section 3.1, a central and defining characteristic of pure mathematics is *proof*. To what extent is it possible to automatically assess mathematical proof? From a theoretical perspective the abstract question is hopeless, since in general it is impossible to verify automatically the correctness of an arbitrary mathematical argument. Despite this potential limitation, the kinds of proof which students write in response to tutorial questions may well be verifiable automatically. This is perhaps the most challenging topic, and one in which progress has been made only in some specialized areas. It is probably as well at this point to recall Babbage's quote at the start of Chapter 3, and state again that CAA appears most useful for the assessment of routine procedures and techniques. In this section we examine the extent to which proofs can currently be automatically assessed. Most of the examples cited here are pilot studies which, at the time of writing at least, have not made the transition to reliable widespread use. Nevertheless, we record them as pointers to work which has been done in this area.

Some proofs in mathematics follow a predictable template; for example, many proofs by induction. This structure makes them amenable to CAA within a relatively fixed item structure, by providing predefined steps. Our first example of this kind is taken from Ruokokoski (2009, Chapter 7) who used STACK to assess proof by mathematical induction of the formulae for $\sum_{r=1}^{n}(br + c)$, where b and c are randomly generated small integers. This was done by providing answer boxes for the various parts of the induction proof. Proofs by induction do have a somewhat predictable structure, and CAA might be an appropriate tool with which to assess this in a supported way. The obvious issue with trying to assess proofs is the difficulty of knowing how many answer boxes to include, and which steps students will take when deriving the induction step. As Ruokokoski (2009) acknowledges, this was not entirely successful.

> This example exercise was used during the fall of 2008, and about 300 students did the exercise. First, we can use many answer fields, but it is not always enough. At present, students are not able to edit the number of answer fields in the induction step. [...] Second, some students used the variable k instead of n [...]. Ruokokoski (2009)

In order to assess the calculation part within the induction step an environment in which steps are controlled by the student would be advantageous, as in the MathXpert system. STACK does not provide this facility, contributing to the difficulties cited by Ruokokoski (2009). In concentrating on calculations constituting the induction step, all the logic has been stripped out. The logical structure of the proof has been provided as steps in the working. In assessing most mathematical proofs it is precisely the logical structure we wish to assess, and so the purely computer algebra based computer aided assessment becomes insufficient. Tools from the automatic theorem proving community are needed. These tools are not available for use in assessment currently. In Section 5.7 we commented on the relatively poor links between the research communities working in automatic theorem proving and computer algebra, even at the research level.

One system more dedicated to the assessment of mathematical proofs is EASy, developed at the University of Muenster. See Gruttmann *et al.* (2008a) and Gruttmann *et al.* (2008b). This consists of a Java applet integrated in a static web page. A screen shot is shown in Figure 9.1. EASy's interface is necessarily rather complex. The central area contains the theorem which has to be proved and the current status of the proof. To the left the student may choose the strategy which will be used—in this case, Hoare Logic. To the right-hand side, the student can select the next rule to be applied and select the position in the last formula of the proof, where this rule should be applied.

EASy has a mathematical core based on conditional term rewriting.

> EASy helps to identify opportunities to solve an exercise by providing proof strategies and applicable rules for each exercise. In addition, EASy supervises the correct use of rules. (Gruttmann *et al.*, 2008a)

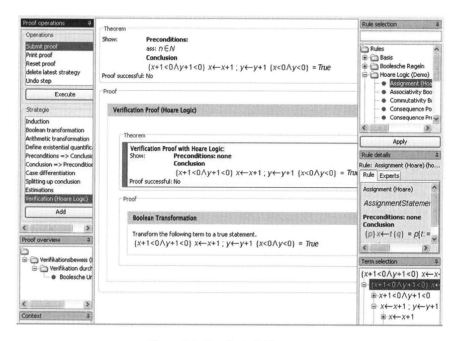

Figure 9.1: Proof in the EASy system.

Just as with MathXpert, the student selects terms, and the system suggests which rules are applicable and then applies the rule chosen by the student.

> Each theorem induces a set of rewrite rules to transform terms. Such a rewrite rule consists of a number of preconditions and a conclusion. The rules engine ensures that the current proof context satisfies the theorem's preconditions. If the engine does not succeed, it cancels the application of the rule. This mechanism guarantees the mathematical correctness of EASy. (Gruttmann et al., 2008a)

An evaluation of the EASy prototype was undertaken at the University of Muenster. About 200 students used the system on a bachelor course 'Data structures and Algorithms' during 2008; see Gruttmann et al. (2008a). Questions in this course included simple proofs such as the formula $\sum_{i=0}^{n} i = \frac{n(n+1)}{2}$. Students in this study were broadly supportive of the system. They reported that they needed more time with EASy, the interface was initially an issue, and '*EASy requires small steps to be executed one by one, which normally could be combined in a paper-based proof*'; Gruttmann et al. (2008a). For example, a Boolean term $x \leq 0 \wedge 0 \geq x$ cannot be transformed by a rule such as $p \geq q \wedge q \geq p$ because the operators $\geq$ and $\leq$ are interchanged. This issue is closely related to that of *recognition*; indeed, it is the automated version.

Subsequent development of the EASy system has moved away from mathematical proof towards the automatic assessment of computer programming exercises. Students are required to write code fragments, and the system then evaluates the properties of this code; see Majchrzak and Usener (2012). Recall that this was precisely the task which motivated the very early work of Hollingsworth (1960). Such evaluations share much in common with CAA of mathematics. In particular, there is a separation of syntax from semantics, and the properties required from the code need to be specified in advance.

> Assessing programming exercises is by no means a trivial task. Different solutions might be semantically equal, and several semantically different solutions might solve a given exercise. (Usener et al., 2012)

Indeed, parallel to the work in mathematical CAA is a whole strand of computer science assessment, with substantially the same objectives. Here, competency-based testing of simple code fragments constitutes the analogy of procedural exercises in algebra and calculus. Students are given randomly generated structured tasks. Indeed, one may safely make the substitution of mathematics for computer science in the following:

> Most of the relevant learning objectives in computer science require intensive involvement and continuous practice. By solving exercises based on lecture contents, students transfer passively consumed information to active knowledge. [...] Hence, continuous formative assessments, such as weekly exercises, are a basic principle of learning and teaching processes in computer science education. (Usener et al., 2012)

Another area of mathematics in which proofs are reasonably predictable is undergraduate real analysis. Such proofs are heavily quantified and often rely on manipulation of real inequalities. Both of these present serious challenges to software. One attempt to develop more flexible environments for such mathematical proofs was the CreaComp project based at the University of Linz, starting in 2004. The project aimed to produce computer-supported learning units for mathematics and these units were developed as Mathematica notebooks, based upon the Theorema automatic theorem proving system, see Buchberger et al. (2006). The aims of this project included assessment only in part. Wider aims were to provide tools in which students explore mathematical concepts using interactive elements and other computer-supported tools.

> Several researchers pointed out, see Sangwin (2006), that CAS can be extremely compressive: a lot of commands produce an output without any intermediate results and details. Thus it leaves the user in a certain unsatisfactory state if he wants to know more about the transformation rules used or how the result of the computation is finally gained. The compressive outputs of the standard CAS computational commands are fine for applications but certainly not appropriate for theory exploration, i.e., for introducing new mathematical notions and methods. (Vajda, 2009)

To overcome this difficulty Theorema offers finer control over the computations carried out, and also supports formal mathematical statements in predicate logic and formal deductions. As an example, Vajda (2009, p. 7) gives the definition of local continuity as follows.

$$\text{Definition}[\text{"cont"}, \text{any}[f; a], \text{with}[a \in \mathbb{R}],$$

$$\text{Continuous}[f; a] \Leftrightarrow \underset{\substack{\epsilon \in \mathbb{R} \\ \epsilon > 0}}{\forall} \underset{\substack{\delta \in \mathbb{R} \\ \delta > 0}}{\exists} \underset{\substack{x \in \mathbb{R} \\ |x-a| < \delta}}{\forall} \left| f[x] - f[a] \right| < \epsilon$$

The input mechanism mirrors very closely the layout and display shown above, and further examples are given in Vajda et al. (2009) together with details of how such definitions are used in automatic deduction. CreaComp is still a work in progress, and it is not clear whether evaluation of the units with students will show that the effort needed to learn how to use the software provides a significant payback. It is clear that serious work to enable students to automate proofs in real analysis is being undertaken, and that ultimately there is a prospect that this will translate into automatic assessment of proofs in this area.

9.3 Semi-automatic marking

A more recent development which addresses both the theoretical and practical difficulties of assessing a complete mathematical argument automatically is *semi-automatic assessment online*. In other disciplines the Internet is used to support assessment processes such as collection/submission of work, distribution to human assessors (perhaps blind/anonymous, and perhaps for double marking), the recording of outcomes and the return of feedback to students. Because this book has concentrated on *automatic assessment* we have had little reason to comment on these processes until this point.

The SAiL-M project ('semiautomatic analysis of individual learning processes in mathematics') is funded by the German Federal Ministry of Education and Research. The project aims to move beyond the assessment of techniques into assessments of problem-solving, justifying and communicating a solution.

> The SAiL-M project aims for a synthesis of automatic and manual assessment. The idea of so-called semi-automatic assessment is that the computer delivers feedback on those aspects of the solution that it can assess, and delegates any further questions to the tutor. (Herding and Schroeder, 2011)

Herding and Schroeder (2011) also discuss related issues such as capture of all student interactions for later review by the tutor. Assessment using semi-automatic feedback in mathematics is discussed further in Bescherer *et al.* (2011), together with some prototype tools. At this stage it is premature to comment on the effectiveness of this approach to assessment.

The obvious drawback of semi-automatic marking is the potential loss of *immediate* feedback. We have seen the extent to which automation is possible, and semi-automatic systems can certainly alert users to technical mistakes. For example, the lack of equivalence between expressions or an incorrect final answer can be highlighted immediately. However, to confirm that a complete solution is really correct will require the attention of a human with an inevitable delay. The immediacy of feedback about technical problems seems potentially very advantageous. The root cause of many errors in students' work are not conceptual misunderstandings, but technical algebraic mistakes. These can be highlighted; indeed, it is possible to envisage an assessment system which will prevent the submission of any work which contains such errors. This would provide students with confidence in their answer and free the teacher from the tedium of pointing out procedural problems to students. This does not address the difficulty of a student who is genuinely stuck and asks for an explanation, but the automatic inclusion of a template of steps might help in some circumstances as it already does in projects such as CALM. The teacher may also have more time to identify and help these individuals directly.

9.4 Standards and interoperability

The lack of a common format into which students can enter a mathematical argument is clearly a serious impediment to progress in this area. This is a significant interface issue. Such a format will need to capture the meaning of expressions, logical connections between steps, supporting rhetorical statements, and perhaps even dynamic graphics linked to other objects. This is a non-trivial task.

9.4 STANDARDS AND INTEROPERABILITY | 171

> In 1876 attention was called to the fact that some nuts cut at one shop would not fit bolts cut at others, and an investigation was made. A set of nuts of different sizes were cut at each of the shops, and were sent to Messrs Pratt & Whitney, who fitted soft plugs, made of Babbit metal, into each of these nuts. [...] By taking at random a plug and a nut of nominally the same diameter, it was found that the one would rarely fit the other. [...] This was the cause of great waste, detention, and expense in making repairs. (Bond, 1887, p. 65)

Creating online learning materials is a genuinely difficult task. The *design* of the questions themselves and subsequent *technical implementation in a particular CAA system* is time-consuming and hence expensive. It requires a number of areas of expertise, in mathematics, education and pedagogy and sometimes computer algebra systems or technical coding skills. Further statistical and educational measurement skills are needed for analysis of the outcomes. The purpose of this book is to discuss these issues. Where a single individual does not possess all these skills a team is needed to create high-quality materials for students. It is therefore natural to seek to reuse and exchange materials which have proved to be effective. To do this specifications for exchange and interoperability are needed which, in the long term, might be recognized as *standards*.

The quotation at the start of this section is taken from the report of the Special Committee appointed to the Franklin Institute on 21 April 1864, to investigate the question of the proper system of screw threads, bold-heads, and nuts. This includes the standards for length and shape.

> Like Diogenes with his lamp, in search of an honest man, this company went to and fro in the land in search of a true inch, a true foot, or a true yard. They procured from different sources what they supposed were the most reliable standards of measurement, and found that none agreed. They had the same standards measured by what were considered the most reliable measuring machines and instruments in the country, and found that no two of these would measure the same standard alike. (Bond, 1887, p. 67)

Given how basic a machine component a nut and bolt is, it might be thought to be a comparatively simple standard to develop. Indeed, they concluded that '*screw-threads shall be formed with straight sides at an angle to each other of* 60°, *having a flat surface at the top and bottom equal to one-eighth of the pitch. These pitches shall be as follows, viz:* [...]'. That is it. Their report was first published in 1887 as (Bond, 1887, p. 65), taking over twenty years to develop. One key difficulty was the ability to reliably measure physical length.

There have been a number of projects with the goal of developing a standard for computer aided assessment, or various components upon which CAA relies, e.g. Lay (2006). However, there has been little real progress. Chapter 5 concerning mathematical notation and Chapter 6 which considers computer algebra discuss the ambiguities and difficulties in ascribing meaning to mathematical symbols in current use and describing their properties. This is needed for assessment, so without standards here corresponding standards for assessment seem a long way off. Indeed, while the current JISC catalogue of standards lists over a hundred related to 'assessment', few specifications are implemented on more than one platform. Standards specify minute details, and by definition do not cover the most innovative features. Bond (1887) is a warning that we should not underestimate the difficulty in developing a 'standard' and of persuading others to use it. The underlying technologies are in such a rapid state of development at the time of writing that a focus on standards at this time seems premature.

9.5 Conclusion

There is a considerable community of practice associated with mathematical computer aided assessment. Many students and teachers routinely make use of CAA. There is growing evidence that students can benefit from undertaking such assessments and the circumstances under which they do so. It is likely that publishers will increasingly provide access to interactive materials and online assessments. The true meaning of 'purchasing a textbook' is likely to move away from ownership of a physical artifact and become closer to gaining access to a suite of online materials. If this is the case, teachers may start to track access and progress in new ways which become less tied to the lock-step classroom. However, we are a considerable way from producing valid assessments of mathematical proofs. Since proofs lie at the heart of mathematical practice, considerable theoretical and practical challenges remain for the future.

BIBLIOGRAPHY

Adams, W. W. and Loustaunau, P. (1994), *An Introduction to Gröbner Bases*, Vol. 3 of *Graduate Studies in Mathematics*, American Mathematical Society.

Anderson, J. R. (1986), *The Architecture of Cognition*, Harvard University Press.

Anderson, J. R., Corbett, A. T., Koedinger, K. R., and Pelletier, R. (1995), Cognitive tutors: Lessons learned, *Journal of the Learning Sciences* **4**(2), 167–207.

Appleby, J., Samuels, P. C., and Jones, T. T. (1997), DIAGNOSYS: a knowledge-based diagnostic test of basic mathematical skills, *Computers in Education* **28**, 113–31.

Ashton, H. S. and Khaled, O. (2009), Scholar intermediate mathematics, *Scottish Mathematical Council Journal* **39**, 45–9.

Ashton, H. S., Beevers, C. E., and Thomas, R. (2008), Can e-assessment become mainstream?, *Proceedings of the CAA conference*.

Ashton, H. S., Beevers, C. E., Korabinski, A. A., and Youngson, M. A. (2005), Investigating the medium effect in school chemistry and college computing national examinations, *British Journal of Educational Technology* **36**(5), 771–87.

Ashton, H. S., Beevers, C. E., Korabinski, A. A., and Youngson, M. A. (2006), Incorporating partial credit in computer-aided assessment of mathematics in secondary education, *British Journal of Educational Technology* **27**(1), 93–119.

Ashton, H. S., Beevers, C. E., Schofield, D. K., and Youngson, M. A. (2004), Informative reports: experiences from the pass-it project, *Proceedings of the 8th International CAA Conference*.

Austin, W. F. (1880), *Mathematical Examination Papers set for Entrance to R. M. A. Woolwich, with Answers*, Edward Stanford, London.

Babbage, C. (1821), Observations on the notation employed in the calculus of functions, *Transactions of the Cambridge Philosophical Society* **I**, 63–76.

Babbage, C. (1827), On the influence of signs in mathematical reasoning, *Transactions of the Cambridge Philosophical Society* **II**, 325–77.

Babbage, C. (1830), On notations, *Edinburgh Encyclopedia* **15**, 394–9.

Babbage, R. H. (1948), The work of Charles Babbage, *Annals of the Computation Laboratory of Harvard University*.

Badger, M. and Sangwin, C. (2011), My equations are the same as yours!: Computer aided assessment using a Gröbner basis approach, *in* A. A. Juan, M. A. Huertas, and C. Steegmann (eds.), *Teaching Mathematics Online: Emergent Technologies and Methodologies*, IGI Global, pp. 259–73.

Badger, M., Sangwin, C. J., and Hawkes, T. O. (2012), Teaching problem-solving in undergraduate mathematics, *Technical Report*, National HE STEM Programme, University of Birmingham.

Barendregt, H. and Cohen, A. M. (2001), Electronic communication of mathematics and the interaction of computer algebra systems and proof assistants, *Journal of Symbolic Computation* **31**(1/2), 3–22.

Barnard, T. (1999), *A Pocket Map of Algebraic Manipulation*, The Mathematical Association.

Beeson, M. (1989), *Computers and Mathematics*, Springer-Verlag, chapter Logic and computation in Mathpert: An expert system for learning mathematics, pp. 202–14.

Beeson, M. (1998), *Computer–Human Interaction in Symbolic Computation*, Springer-Verlag, chapter Design Principles of Mathpert: Software to Support Education in Algebra and Calculus, pp. 89–115.

Beeson, M. (2003), *Alan Turing: Life and Legacy of a Great Thinker*, Springer, chapter The Mechanization of Mathematics, pp. 77–134.

Beevers, C. E., Cherry, B. S. G., Foster, M. G., and McGuire, G. R. M. (1991), *Software Tools for Computer Aided Learning in Mathematics*, Avebury Technical.

Beevers, C. E., McGuire, G. R., Stirling, G., and Wild, D. G. (1995), Mathematical ability assessed by computer, *Journal for Computers and Education* **25**, 123–32.

Bennett, R. E., Steffen, M., Singley, M. K., Morley, M., and Jacquemin, D. (1997), Evaluating an automatically scorable, open-ended response time for measuring mathematics reasoning in computer-adaptive tests, *Journal of Educational Measurement* **34**(2), 162–76.

Bescherer, C., Herding, D., Kortenkamp, U., Müller, W., and Zimmermann, M. (2011), *Intelligent and Adaptive Learning Systems: Technology Enhanced Support for Learners and Teachers*, IGI Global, chapter E-Learning Tools with Intelligent Assessment and Feedback, pp. 151–63.

Blåfield, L. (2009), *Matematiikan verkko-opetus osana perusopetuksen kehittämistä teknillisessä korkeakoulussa*, Master's thesis, University of Helsinki.

Bloom, B. S., Engelhart, M. D., Furst, E. J., Hill, W. H., and Krathwohl, D. R. (1956), *Taxonomy of Educational Objectives: Cognitive Domain*, McKay, New York.

Bond, G. E. (1887), *Standards of Length and their Practical Application*, The Pratt and Whitney Company.

Bradford, R., Davenport, J. H., and Sangwin, C. J. (2009), A Comparison of Equality in Computer Algebra and Correctness in Mathematical Pedagogy, *Proceedings of Calculemus*, number 5625 in *Lecture Notes in Artificial Intelligence*, pp. 75–89.

Bradford, R., Davenport, J. H., and Sangwin, C. J. (2010), A comparison of equality in computer algebra and correctness in mathematical pedagogy (ii), *The International Journal for Technology in Mathematics Education* **17**(2), 93–8.

Brousseau, G. (1997), *Theory of Didactical Situations in Mathematics: Didactiques des mathématiques, 1970–1990*, Kluwer. N. Balacheff, M. Cooper, R. Sutherland, and V. Warfield (trans.).

Brown, M. (1974), Some thoughts on the use of computer symbols in mathematics, *The Mathematical Gazette* **58**(404), 78–9.

Brown, W. S. (1969), Rational exponential expressions and a conjecture concerning π and e, *American Mathematical Monthly* **76**(1), 28–34.

Buchberger, B. (1990), Should students learn integration rules?, *ACM SIGSAM Bulletin* **24**(1), 10–17.

Buchberger, B., Crăciun, A., Jebelean, T., Kovács, L., Kutsia, T., Nakagawa, K., Piroi, F., Popov, N., Robu, J., Rosenkranz, M., and Windsteiger, W. (2006), Theorema: Towards computer-aided mathematical theory exploration, *Journal of Applied Logic* **4**(4), 470–504.

Bundy, A. (1983), *The Computer Modelling of Mathematical Reasoning*, Academic Press.

Burkhardt, H. and Swan, M. (2012), Designing assessment of performance in mathematics, *Educational Designer: Journal of the International Society for Design and Development in Education*.

Burn, R. P. (1987), *Groups: A Path to Geometry*, Cambridge University Press.

Burn, R. P. (1996), *A Pathway into Number Theory*, Cambridge University Press.
Burn, R. P. (2000), *Numbers and Functions: Steps to Analysis*, Cambridge University Press.
Burton, R. R. (1982), Diagnosing bugs in a simple procedural skill, in D. Sleeman and J. S. Brown (eds.), *Intelligent Tutoring Systems*, Academic Press, chapter 8, pp. 157–83.
Butcher, P. G. and Jordan, S. E. (2010), A comparison of human and computer marking of short free-text student responses, *Computers and Education* **55**(2), 489–99.
Buxton, H. W. and Hyman, A. (1987), *Memoir of the Life and Labours of the Late Charles Babbage*, MIT Press.
Cajori, F. (1928), *A History of Mathematical Notations*, Open Court.
Cardano, G. (1993), *Ars Magna, or, The Rules of Algebra*, Dover, New York.
Carette, J. (2004), Understanding expression simplification, in J. Gutierrez (ed.), *Proceedings of ISSAC 2004*, pp. 72–9.
Caviness, B. F. (1970), On canonical forms and simplification, *Journal of the ACM* **17**(2), 385–96.
Cerval-Peña, E. R. (2008), *Automated computer-aided formative assessment with ordinary differential equations*, Masters thesis, University of Birmingham.
Chatterji, M. (2003), *Designing and using tools for educational measurement*, Allyn and Bacon.
Chrystal, G. (1893), *Algebra: An Elementary Text-book for the Higher Classes of Secondary Schools and for Colleges*, Vol. 1, third edn., Adam and Charles Black, London and Edinburgh.
Chuquet, N. (1880), Le Triparty en la Science des Nombres, éd. A. Marre, *Bullettino di Bibliografia e di Storia delle scienze matematiche et fisiche* **13**, 555–659.
Cockcroft, W. H. (1982), *Mathematics Counts*, Her Majesty's Stationery Office.
Cooper, B. and Dunne, M. (2000), *Assessing Children's Mathematical Knowledge: Social Class, Sex and Problem-Solving*, Open University Press.
Corless, R. M., Gonnet, G. H., Hare, D. E. G., Jeffrey, D. J., and Knuth, D. E. (1996), On the LambertW function, *Advances in Computational mathematics* **5**(1), 329–59.
Dahlberg, R. P. and Housman, R. P. (1997), Facilitating learning events through example generation, *Educational Studies in Mathematics* **33**(3), 283–99.
Davenport, J. H. (2002), Equality in computer algebra and beyond, *Journal of Symbolic Computation* **34**, 259–70.
Davenport, J. H. and Carette, J. (2009), The sparsity challenges, *in* S. Watt (ed.), *Proceedings of the 11th International Symposium on Symbolic and Numeric Algorithms for Scientific Computing (SYNASC 2009)*, IEEE Computer Society, pp. 3–7. http://opus.bath.ac.uk/17324/.
Davenport, J. H., Siret, Y., and Tournier, E. (1993), *Computer Algebra: Systems and Algorithms for Algebraic Computation*, Academic Press Professional.
Dedic, H., Rosenfield, S., and Ivanov, I. (2007), Just computer aided instruction is not enough: Combining webwork with in-class interactive sessions increases achievement and perseverance of social science calculus students, Preprint.
Descartes, R. (1952), *The Geometry of René Descartes; Translated from the French and Latin by D. E. Smith and M. L. Latham*, Open Court.
Edwards, C. H. (1979), *The Historical Development of the Calculus*, Springer-Verlag.
Ericsson, K. A., Krampe, R., and Tesch-Römer, C. (1993), The role of deliberate practice in the acquisition of expert performance, *Psychological Review* **100**(3), 363–406.
Euler, L. (1988), *Introduction to Analysis of the Infinite*, Vol. I, Springer. Translated by J. Blanton, from the Latin *Introductio in Analysin Infinitorum*, 1748.

Euler, L. (1990), *Introduction to Analysis of the Infinite*, Vol. II, Springer. Translated by J. Blanton, from the Latin *Introductio in Analysin Infinitorum, 1748*.

Euler, L. (2006), *Elements of Algebra*, Tarquin Publications. ISBN 978-1-89961-873-6.

Fenichel, R. R. (1966), *An On-line System for Algebraic Manipulation*, Phd thesis, Harvard Graduate School of Arts and Sciences.

Fitch, J. (1973), On algebraic simplification, *Computer Journal* **16**(1), 23–27.

Foster, B., Perfect, C., and Youd, A. (2012), A completely client-side approach to e-assessment and e-learning of mathematics and statistics, *International Journal of eAssessment* **2**(2), 1–12.

Fujimoto, M. and Suzuki, M. (2003), *Infty Editor: A Mathematics Typesetting Tool with a Handwriting Interface and a Graphical Front-end to OpenXM Servers*, Vol. 1335 of *Computer Algebra: Algorithms, Implementations and Applications*, RIMS Kokyuroku, pp. 217–26.

Furber, S. (2012), Shut down or restart? the way forward for computing in UK schools, *Technical report*, The Royal Society.

Gage, M., Pizer, A., and Roth, V. (2002), WeBWorK: Generating, delivering, and checking math homework via the Internet, Proc. ICTM2 international congress for teaching of mathematics at the undergraduate level, http://www.math.uoc.gr/ictm2/Proceedings/pap189.pdf.

Gardiner, A. (2003), *Understanding Infinity: The Mathematics of Infinite Processes*, Dover.

Gardiner, T. (2006), Beyond the soup kitchen: Thoughts on revising the mathematics 'strategies/ frameworks' for England, *International Journal for Mathematics Teaching and Learning*. ISSN 1473–0111.

Gattengo, C. (1988), *The Science of Education, Part 2B: The Awareness of Mathematization*, Educational Solutions, New York.

Gelbaum, B. R. and Olmsted, J. M. H. (1964), *Counterexamples in Analysis*, Holden-Day.

Gerofsky, S. G. (1999), *The Word Problem as Genre in Mathematics Education*, PhD thesis, Simon Fraser University.

Gibbs, G. (1999), Using assessment strategically to change the way students learn, *Assessment Matters in Higher Education: Choosing and Using Diverse Approaches*, Society for Research into Higher Education & Open University Press, pp. 41–53.

Gibbs, G. and Simpson, C. (2004), Conditions under which assessment supports students' learning, *Learning and Teaching in Higher Education* **1**, 3–32.

Grabmeier, J., Kaltofen, E., and Weispfenning, V. (2003), *Computer Algebra Handbook*, Springer.

Gray, E. M. and Tall, D. O. (1994), Duality, ambiguity and flexibility: A proceptual view of simple arithmetic, *Journal for Research in Mathematics Education* **26**(2), 115–41.

Greenhow, M. and Gill, M. (2008), Computer-aided assessment in mechanics: question design and test evaluation, *Teaching Mathematics and its Applications* **26**(3), 124–33. doi:10.1093/teamat/hrm006.

Gruttmann, S., Böhm, D., and Kuchen, H. (2008a), E-assessment of mathematical proofs: chances and challenges for students and tutors, *Proceedings of the 2008 International Conference on Computer Science and Software Engineering*, IEEE Computer Society, pp. 612–5.

Gruttmann, S., Böhm, D., and Kuchen, H. (2008b), An e-assessment system for mathematical proofs, *Proceedings of the IASTED International Conference on Computer and Advanced Technology in Education*, ACTA Press, pp. 120–5.

Hadley, J. and Singmaster, D. (1992), Problems to sharpen the young, *The Mathematical Gazette* **76**(475), 102–26.

Harding, R. and Quiney, D. (1996), Mathwise and the UKMCC, *Active Learning* **4**, 53–7.

Hardy, G. H. and Wright, E. M. (1960), *An Introduction to the Theory of Numbers*, 4th edn, Oxford University Press.

Harjula, M. (2008), *Mathematics exercise system with automatic assessment*, Masters thesis, Helsinki University of Technology.

Hassmén, P. and Hunt, D. P. (1994), Human self-assessment in multiple choice, *Journal of Educational Measurement* **31**(2), 149–60.

Havola, L. (2010), Improving the teaching of engineering mathematics: a research plan and work in-process report, *In proceedings of the Joint International IGIP-SEFI Annual Conference 2010, Trnava, Slovakia*.

Havola, L. (2011), *Tutkimus suuntaamassa 2010-luvun matemaattisten aineiden opetusta*, Tampereen yliopistopaino Oy – Juvenes Print, chapter New engineering students' learning styles and basic skills in mathematics, pp. 118–31.

Hayes, B. (2009), Writing math on the web, *American Scientist* **97**(2), 98–102.

Heck, A. (2003), *Introduction to Maple*, 3rd edn, Springer-Verlag.

Heeren, B. and Jeuring, J. (2009), Canonical forms in interactive exercise assistants, in J. Carette, L. Dixon, C. Sacerdoti Coen and S. M. Watt (eds.), *Intelligent Computer Mathematics: Proceedings of the 8th International Conference on Mathematical Knowledge Management*, Vol. 5625, Springer-Verlag, pp. 325–40.

Heeren, B., Jeuring, J., and Gerdes, A. (2010), Specifying rewrite strategies for interactive exercises, *Mathematics in computer science* **3**(3), 349–70.

Hehner, C. R. (2004), from Boolean algebra to unified algebra, *The Mathematical Intelligencer* **26**(2), 3–9.

Hendrix, G. (1961), Learning by discovery, *The Mathematics Teacher* **54**, 290–9.

Herding, D. and Schroeder, U. (2011), Using capture and replay for semi-automatic assessment, *Proceedings of CAA 2011*.

Hérigone, P. (1634), *Cursus mathematicus, nova, brevi, et clara methodo demonstratus*, Paris.

Hewitt, D. (1996), Mathematical fluency: the nature of practice and the role of subordination, *For the learning of mathematics* **16**(2), 28–35.

Hobbes, T. (1656), *Six lessons to the Professors of Mathematiques, one of Geometry, the other of Astronomy: in the Chaires set up by Sir Henry Savile in the University of Oxford*, Oxford, London.

Hoffmann, B. (1962), *The Tyranny of Testing*, Crowell-Collier.

Hollingsworth, J. (1960), Automatic graders for programming classes, *Communications of the ACM* **3**(10), 528–9.

Horsley, S. (1782), *Isaaci Newtoni Opera*, Joannes Nichols, London.

Johnson, C. (2011), Nice cubics, *The Mathematical Gazette* **95**(533), 273–9.

Jordan, S. (2011), Using interactive computer-based assessment to support beginning distance learners of science, *Open Learning* **26**(2), 147–64.

Jordan, S. and Butcher, P. (2010), *Physics Community and Cooperation: Selected Contributions from the GIREP-EPEC and PHEC 2009 International Conference*, Lulu/The Centre for Interdisciplinary Science, chapter Using e-assessment to support distance learners of science, pp. 202–16.

Jordan, S., Jordan, H., and Jordan, R. (2011), Same but different, but is it fair? An analysis of the use of variants of interactive computer-marked questions, *Proceedings of International Computer Assisted Assessment Conference*.

Kehoe, E. (2010), AMS homework software survey, *Notices of the American Mathematical Society* **57**(6), 753–7.

Kerber, M. and Pollet, M. (2007), Informal and formal representations in mathematics, *Studies in Logic, Grammar and Rhetoric* **10**(23), 75–95.

Kirshner, D. (1989), The visual syntax of algebra, *Journal for Research in Mathematics Education* **20**(3), 274–287.

Kirshner, D. and Awtry, T. (2004), Visual salience of algebraic transformations, *Journal for Research in Mathematics Education* **35**, 224–57.

Klai, S., Kolokolnikov, T., and Van den Bergh, N. (2000), Using Maple and the web to grade mathematics tests, *Proceedings of the International Workshop on Advanced Learning Technologies, Palmerston North, New Zealand, 4–6 December*.

Kluger, A. N. and DeNisi, A. (1996), Effects of feedback intervention on performance: A historical review, a meta-analysis, and a preliminary feedback intervention theory., *Psychological Bulletin* **119**(2), 254–84.

Knuth, D. E. (1969), *The Art of Computer Programming*, Vol. 2, Addison-Wesley, Reading, MA.

Knuth, D. E. (1979), *TEX and METAFONT: New Directions in Typesetting*, American Mathematical Society.

Kolb, D. A. (1984), *Experiential Learning: Experience as the Source of Learning and Development*, Prentice Hall.

Krause, E. F. (1975), *Taxicab Geometry: An Adventure in Non-Euclidean Geometry*, Addison-Wesley.

Lakatos, I. (1976), *Proofs and Refutations*, Cambridge University Press.

Landau, S. (1992), Simplification of nested radicals, *SIAM Journal on Computing* **21**(1), 85–110.

Laurillard, D. (2002), *Rethinking University Teaching, A Conversational Framework for the Effective Use of Learning Technology*, 2 edn., Routledge Falmer.

Lay, S. (2006), *IMS Question and Test v2.1 Public Draft 2*, IMS Global Learning Consortium.

Leder, G. C., Rowley, G., and Brew, C. (1999), *International Comparisons in Mathematics: The State of the Art*, Falmer Press, chapter Gender differences in mathematics achievement: Here today and gone tomorrow?, pp. 213–24.

Libbrecht, P. (2010), *Intelligent Computer Mathematics*, Vol. 6167, Springer, chapter Notations Around the World: Census and Exploitation, pp. 398–401.

Lowe, J. (2001), Computer-based education: Is it a panacea?, *Journal of Research on Technology in Education* **34**(2), 163–71.

Majander, H. (2010), *Tietokoneavusteinen arviointi kurssilla diskreetin matematiikan perusteet*, Master's thesis, University of Helsinki.

Majander, H. and Rasila, A. (2011), *Tutkimus suuntaamassa 2010-luvun matemaattisten aineiden opetusta*, Tampereen yliopistopaino Oy – Juvenes Print, chapter Experiences of continuous formative assessment in engineering mathematics, pp. 197–214.

Majchrzak, T. A. and Usener, C. A. (2012), Evaluating e-assessment for exercises that require higher-order cognitive skills, *Proceedings of 45th Hawaii International Conference on System Sciences*, pp. 48–57.

Marriott, J., N., D., and L., G. (2009), Teaching, learning and assessing statistical problem solving, *Journal of Statistics Education* **17**(1) (online).

Mason, J. (2001), On the use and abuse of word problems for moving from arithmetic to algebra, The University of Melbourne, pp. 430–7.

Mason, J. and Johnston-Wilder, S. (2004), *Fundamental Constructs in Mathematics Education*, Routledge Falmer.
Mason, J. and Klymchuk, S. (2009), *Using Counter-examples in Calculus*, Imperial College Press.
Mason, J. and Pimm, D. (1984), Generic examples: seeing the general in the particular, *Educational Studies in Mathematics* **15**, 277–89.
Mason, J. and Watson, A. (2001), Getting students to create boundary examples, *MSOR Connections* **1**(1), 9–11. http://ltsn.mathstore.ac.uk/ (viewed August 2002).
Matiyasevich, Y. (1993), *Hilbert's Tenth Problem*, MIT, Cambridge.
Matz, M. (1982), Towards a process model for high school algebra errors, in D. Sleeman and J. S. Brown (eds), *Intelligent Tutoring Systems*, Academic Press, chapter 2, pp. 25–50.
Maxwell, E. A. (1959), *Fallacies in Mathematics*, Cambridge University Press.
Mayer, R. E. (1981), Frequency norms and structural analysis of algebra story problems into families, categories, and templates, *Instructional Science* **10**, 135–75.
McAlister, M. (2005), Formative assessment in mathematics using the 'ALTA' system, *Technical Report*, Stranmillis University College, Belfast.
McGuire, G. R., Youngson, M. A., Korabinski, A. A., and McMillan, D. (2002), Partial credit in mathematics exams: a comparison of traditional and CAA exams, *Proceedings 6th International CAA Conference, Loughborough University*.
Michener, E. R. (1978), Understanding understanding mathematics, *Cognitive Science* **2**, 361–81.
Miller, B. R. (1995), An expression formatter for Macsyma. http://math.nist.gov/~BMiller/computer-algebra/.
Moore, R. C. (1994), Making the transition to formal proof, *Educational Studies in Mathematics* **27**, 249–66.
Morgan, C. (1998), *Writing Mathematically: The Discourse of Investigation*, Fulmer, London.
Morgan, C., Tsatsaroni, A., and Lerman, S. (2002), Mathematics teachers' positions and practices in discourses of assessment, *British Journal of Sociology of Education* **23**(3), 445–61.
Moses, J. (1971), Algebraic simplification: A guide for the perplexed, *Communications of the ACM* **14**(8), 527–37.
Naismith, L. and Sangwin, C. J. (2004), Computer algebra based assessment of mathematics online, *Proceedings of the 8th CAA Conference 2004, 6th and 7th July, The University of Loughborough, UK*.
Nakamura, Y. (2010), *The STACK E-Learning and Assessment System for mathematics, science and Engineering Education through Moodle*, Tokyo Denki University Press, Preface, pp. vi–vii. In Japanese. ISBN 978-4-501-54820-9.
Nicaud, J. F. and Bouhineau, D. (2008), Natural Editing of Algebraic Expressions, *Les Cahiers Leibniz*.
Nicaud, J. F., Bouhineau, D., and Chaachoua, H. (2004), Mixing microworlds and CAS features in building computer systems that help students learn algebra, *International Journal of Computers for Mathematical Learning* **9**(2), 169–211.
Niss, M. (ed.) (1993), *Investigations into Assessment in Mathematics Education: an ICMI Study*, The Netherlands: Kluwer Academic.
Northrop, E. P. (1945), *Riddles in Mathematics: A Book of Paradoxes*, The English Universities Press.
Noss, R. and Hoyles, C. (1996), *Windows on Mathematical Meanings: Learning Cultures and Computers*, Springer.
Nunn, T. P. (1911), The aim and methods of school algebra. 1. the aim of algebra teaching, *The Mathematical Gazette* **6**(95), 167–72.

Oettinger, A. G. (1974), *Run, Computer, Run: Mythology of Educational Innovation*, Harvard University Press.
Oughtred, W. (1652), *Clavis Mathematicæ*, Oxford.
Papert, S. (1980), *Mindstorms: Children, Computers and Powerful Ideas*, Harper Collins.
Parker, J. (2004), *R. L. Moore: Mathematician and Teacher*, Mathematical Association of America.
Pitcher, N. (1997), Educational software in mathematics: developing and using a mathwise module, *International Journal of Mathematical Education in Science and Technology* **29**(5), 709–20.
Pointon, A. and Sangwin, C. J. (2003), An analysis of undergraduate core material in the light of hand held computer algebra systems, *International Journal of Mathematical Education in Science and Technology* **34**(5), 671–86.
Polya, G. (1962), *Mathematical Discovery: On Understanding, Learning, and Teaching Problem Solving*, Wiley.
Polya, G. (1973), *How to Solve It*, Princeton University Press.
Prank, R. (2008), Random generation of expressions in problem solving environment t-algebra, *Journal of Computers in Mathematics and Science Teaching*.
Prank, R. (2011), What toolbox is necessary for building exercise environments for algebraic transformations, *The Electronic Journal of Mathematics and Technology*.
Quigley, M. T. (1988), *Computer Tutoring in Mathematics Education Using Artificial Intelligence Tools*, PhD thesis, University of Birmingham.
Rasila, A., Harjula, M., and Zenger, K. (2007), Automatic assessment of mathematics exercises: Experiences and future prospects, *ReflekTori 2007: Symposium of Engineering Education*, Helsinki University of Technology, Finland, Teaching and Learning Development Unit, http://www.dipoli.tkk.fi/ok, pp. 70–80.
Rasila, A., Havola, L., H., M., and Malinen, J. (2010), Automatic assessment in engineering mathematics: evaluation of the impact, *ReflekTori 2010: Symposium of Engineering Education*, Aalto University, Finland, Teaching and Learning Development Unit, http://www.dipoli.tkk.fi/ok.
Recorde, R. (1557), *The Whetstone of Witte*, London, I. Kyngston.
Renteln, P. and Dundes, A. (2005), Foolproof: A sampling of mathematical folk humor, *Notices of the American Mathematical Society* **52**(1), 24–34.
Richardson, D. (1966), *Solvable and Unsolvable Problems Involving Elementary Functions of a Real Variable*, PhD thesis, University of Bristol.
Ridgeway, J., McCusker, S., and Pead, D. (2004), Literature review of E–assessment, *Futurelab Series 10*, Futurelab. ISBN: 0-9544695-8-5.
Roach, M., Blackmore, P., and Dempster, J. (2001), Supporting high-level learning through research-based methods, *Innovations in Education and Training International*. http://www.telri.ac.uk/ (viewed June 2008).
Robson, E. (2001), Neither Sherlock Holmes nor Babylon: A reassessment of Plimpton 322, *Historia Mathematica* **28**(3), 167–206.
Robson, E. (2008), *Mathematics in Ancient Iraq*, Princeton University Press.
Roos, B. and Hamilton, D. (2005), Formative assessment: a cybernetic viewpoint, *Assessment in Education: Principles, Policy & Practice* **12**(1), 7–20.
Ross, S. M., Jordan, S. E., and Butcher, P. G. (2006), *Innovative assessment in Higher Education*, Routledge, chapter Online instantaneous and targeted feedback for remote learners, pp. 123–31.

Roth, V., Ivanchenko, V., and Record, N. (2008), Evaluating student response to webwork, a web-based homework delivery and grading system, *Computers and Education* **50**(4), 1462–82.

Ruokokoski, J. (2009), *Automatic assessment in university-level mathematics*, Masters thesis, Helsinki University of Technology.

Sadler, D. R. (1998), Formative assessment: revisiting the territory, *Assessment in Education: Principles, Policy & Practice* **5**(1), 77–84.

Sangwin, C., Cazes, C., Lee, A., and Wong, K. L. (2009), Micro-level automatic assessment supported by digital technologies, *Mathematics Education and Technology: Rethinking the Terrain*, Vol. 13 of *New ICMI Study Series*, Springer, pp. 227–50. DOI: 10.1007/978-1-4419-0146-0.

Sangwin, C. J. (2003a), Providing feedback to students' assignments, *Proceedings of British Society for Research into Learning Mathematics, November 15, The University of Birmingham, Birmingham, UK*, pp. 55–60.

Sangwin, C. J. (2003b), New opportunities for encouraging higher-level mathematical learning by creative use of emerging computer aided assessment, *International Journal of Mathematical Education in Science and Technology* **34**(6), 813–29.

Sangwin, C. J. (2005), On building polynomials, *The Mathematical Gazette* **89**(516), 441–51.

Sangwin, C. J. (2006), Computer Algebra through a Proceptual Lens, in A. Simpson (ed.), *Retirement as Process and Concept: a Festschrift for Eddie Gray and David Tall*, pp. 215–22.

Sangwin, C. J. (2010), Who uses STACK? A report on the use of the STACK CAA system, *Technical report*, The Maths, Stats and OR Network, School of Mathematics, University of Birmingham.

Sangwin, C. J. (2011a), Limit-free derivatives, *The Mathematical Gazette* (534), 469–82.

Sangwin, C. J. (2011b), Modelling the journey from elementary word problems to mathematical research, *Notices of the American Mathematical Society* **58**(10), 1436–45.

Sangwin, C. J. (2012), The dragmath equation editor, *MSOR Connections*.

Sangwin, C. J. and Grove, M. J. (2006), STACK: Addressing the needs of the 'neglected learners', *Proceedings of the First WebALT Conference and Exhibition January 5–6, Technical University of Eindhoven, Netherlands*, Oy WebALT Inc, University of Helsinki, ISBN 952-99666-0-1, pp. 81–95.

Sangwin, C. J. and Ramsden, P. (2007), Linear syntax for communicating elementary mathematics, *Journal of Symbolic Computation* **42**(9), 902–34. DOI: 10.1016/j.jsc.2007.07.002.

Shute, V. J. (2007), *The Future of Assessment: Shaping Teaching and Learning.*, Taylor and Francis Group, chapter Tensions, trends, tools, and technologies: Time for an educational sea change., pp. 139–87.

Sleeman, D. and Brown, J. S. (eds) (1982), *Intelligent Tutoring Systems*, Academic Press.

Smith, G., Wood, L., Coupland, M., and Stephenson, B. (1996), Constructing mathematical examinations to assess a range of knowledge and skills, *International Journal of Mathematics Education in Science and Technology* **27**(1), 65–77.

Stedall, J. A. (2000), Ariadne's Thread: The Life and Times of Oughtred's Clavis, *Annals of Science* **57**(1), 27–60.

Stedall, J. A. (2002), *A Discourse Concerning Algebra: English Algebra to 1685*, Oxford University Press, USA.

Steel, A. (2002), A new scheme for computing with algebraically closed fields, *Lecture Notes In Computer Science* pp. 491–505.

Steele, J. D. (2003), Setting linear algebra problems, Preprint, University of New South Wales, PM 97/2. http://web.maths.unsw.edu.au/~jds/Papers/linalg.pdf.

Stoutemyer, D. R. (1991), Crimes and misdemeanors in the computer algebra trade, *Notices of the American Mathematical Society* **38**(7), 778–85.
Strickland, N. (2002), Alice interactive mathematics, *MSOR Connections* **2**(1), 27–30. http://ltsn.mathstore.ac.uk/newsletter/feb2002/pdf/aim.pdf (viewed December 2002).
Suppes, P. (1967), Some theoretical models for mathematics teaching, *Journal of research and development in education* **1**, 5–22.
Swift, J. (1726), *Gulliver's Travels*, Motte, B.
Tuckey, C. O. (1904), *Examples in Algebra*, Bell & Sons, London.
Tuckey, C. O. (1934), *The Teaching of Algebra in Schools*, A Report for the Mathematical Association, G. Bell & Sons.
Urban, P., Owen, J., Martin, D., Haese, R., Haese, S., and Bruce, M. (2005), *Mathematics HL (core)*, Mathematics for the International Student, second edn., Haese and Harris Publications.
Usener, C. A., Majchrzak, T. A., and Kuchen, H. (2012), E-assessment and software testing, *Interactive Technology and Smart Education* **9**(1), 45–56.
Vajda, R. (2009), An e-learning environment for elementary analysis: combining computer algebra, graphics and automated reasoning, *Teaching Mathematics and Computer Science* **7**(1), 13–34.
Vajda, R., Jebelean, T., and Buchberger, B. (2009), Combining logical and algebraic techniques for natural style proving in elementary analysis, *Mathematics and Computers in Simulation* **79**(8), 2310–16. Special Issue on Nonstandard Applications of Computer Algebra.
Viete, F. (1636), *Algebre de Viete, d'une methode nouvelle, claire, et facile*, L. Boulenger, Paris.
Wall, H. S. (1969), *Creative Mathematics*, University of Texas Press.
Wallis, J. (2004), *The Arithmetic of Infinitesimals*, Springer.
Watson, A. and Mason, J. (2002a), Extending example space as a learning/teaching strategy in mathematics, in A. D. Cockburn and E. Nardi (eds.), *Proceedings of the Annual Conference of the International Group for the Psychology of Mathematics Education (PME26, Norwich, United Kingdom)*, Vol. 4, pp. 378–85.
Watson, A. and Mason, J. (2002b), Student-generated examples in the learning of mathematics, *Canadian Journal for Science, Mathematics and Technology Education* **2**(2), 237–49.
Watson, A. and Mason, J. (2006), Seeing an exercise as a single mathematical object: Using variation to structure sense-making, *Mathematical Thinking and Learning* **8**(2), 91–111.
Wester, M. (1999), *Computer Algebra Systems: A Practical Guide*, Wiley.
Wiener, N. (1968), *The Human Use of Human Beings: Cybernetics and Society*, Sphere Books, London.
Wild, I. (2009), *Moodle 1.9 Math*, Packt Publishing.
Wiliam, D. and Black, P. J. (1996), Meanings and consequences: a basis for distinguishing formative and summative functions of assessment?, *British Educational Research Journal* **22**(5), 537–48.
Xiao, G. (2001), Wims: an interactive mathematics server, *Journal of online Mathematics and its applications*. http://www.joma.org/.
Yates, R. C. (1949), *Geometrical Tools: A Mathematical Sketch and Model Book*, Educational Publishers Incorporated, Saint Louis.

INDEX

absolute value, 75, 85
adaptive tests, 6, 10, 18, 42, 98, 106, 107, 113, 134, 140–147, 157
AiM, 8, 38, 40, 66–67, 74, 75, 102–104
Alcuin of York, 28
Aleks, 160
algebra, 20
   cossic, 55
algebrators, 150
algorithm
   Euclidean, 26, 96
   standard, 20, 25–27, 40, 47, 97, 106, 129, 145
ambiguity, 83, 86, 131, *see* syntax, ambiguity
American Mathematical Society, 151
analysis
   complex, 152
   numerical, 75
   real, 6, 51, 83, 169
answer note, 14, 99, 103, 113
answer test, 113
Aplusix, 84, 145, 157–159, 163
assessment
   criteria-referenced, 23
   definition of, 21
   design of, 25–33, 35, 37–52, 142, 149
   diagnostic, 22, 130, 140
   difficulty, 46
   equivalent, 38
   evaluative, 22, 23, 113
   follow-through, 6, 109, 110, 134, 137
   formative, 22, 113, 148, 154
   high stakes, 10, 24, 38, 124, 135–137
   ipsative, 23
   norm-referenced, 23
   objective, 2–4, 23, 74
   peer, 21
   practicality, 23, 31, 35, 127
   purpose of, 22–23
   random, 10, 13, 14, 35, 38–44, 74, 98, 99, 103, 105, 106, 112–114, 131, 132, 142, 148, 159, 166, 168
   reliability, 23, 31
   self, 21
   semi-automatic, 169–170
   stakes, 22, 25, 54
   summative, 22, 113, 119
   taxonomies, 46–49
   traditional, 5–6, 35, 38, 70, 120, 123, 136–137, 154, 168
   validity, 3, 23, 32, 33, 39, 107, 138, 139, 146, 156
authoring, 104, 105, 107–111, 138, 149
automatic theorem proving, 157, 162, 164, 169
Axiom, 61, 64, 74, 76

Babbage, Charles, 1, 19, 55, 58, 86
Babylonian mathematics, 19, 37
bags, 89
Buchberger's algorithm, 96
buggy rules, *see* production rules, buggy

CABLE, 74
calculus, 20, 142, 154
CALM, 68, 75, 129–132, 139, 140, 149, 170
canonical form, *see* form, canonical
CAPA, 150
Cardano, Gerolamo, 85
Carnegie, 160
CAS, *see* computer algebra
chemistry, 137
Chuquet, Nicolas, 55
Cognitive tutors, 146–147
Collins, John, 55
complex analysis, *see* analysis, complex

computer algebra, 73–101, 155
   side conditions, 83, 100, 155
computer algebra system, *see* computer algebra
confidence testing, 3
constructivism, 34
continuity, 96, 169
conventions, 5, 27, 56–60
CreaComp, 169
cubic, *see* polynomial, cubic
CUE, 129
curriculum, 24

Derive, 61, 62, 74, 76
Descartes, René, 55
Diagnosys, 140–145
didactic contract, 43
didactic tension, 30
differentiability, 96
differential equations, 31, 42, 48
dimensions of possible variation, 41
distracters, 3
DragMath, 68–69, 111

EASy, 167–168
equality, 61
equations, 73, 89, 94
equivalence, 88–91
   algebraic, 75–76, 100, 129, 140, 153
   reasoning by, 84, 158
   systems of equations, 94
   theoretical limitations, *see* undecidability
Eratosthenes, sieve of, 26
Euclid, 26
Euler, Leonhard, 56, 86
evaluation, *see* assessment, evaluative
examinations, 5, 21, 124–125, 135–137
example generation, 4, 45, 49, 75, 102

example space, 41
expert system, 98, 140, 146
exponential function, 73, 85
expression tree, *see* syntax, expression trees

Farey sequences, 98
feedback, 4, 13–15, 21, 22, 33–35, 43, 49, 51, 66, 70, 71, 98, 99, 104–106, 108, 111, 113, 124, 128, 130, 131, 137, 146, 157, 158, 170
  definition of, 14, 34
  delayed, 170
  effectiveness, 33
  formative, 5, 11, 15–17, 34, 102, 106
  general, *see* solution, worked
  immediate, 122, 137, 151, 159, 160, 170
  progressive, 139
folklore, 27
form
  canonical, 5, 41, 84, 91
  expanded, 94
  factored, 92–94
  partial fraction, 94

Gaussian elimination, 96
gender, 3
GeoGebra, 50, 70–71, 105, 114, 115
geometry, 57
Gröbner basis, *see* polynomial, Gröobner basis
graph, 44, 50, 134, 154, 170
graph theory, 42
group theory, 48

Handel, George, 26
Harriot, Thomas, 55
Hawthorn effect, 149
Hérigone, Pierre, 55
Herschel, John, 1
higher-level skills, 24
hints, 147, 154
Hume, James, 55

impersonation, 10, 32, 38
induction, *see* proof by induction
inequalities, 5, 65, 75, 89, 96, 101–169

integration, 3, 9, 14, 27, 31, 128, 154, 165
  constant of, 11
intelligent tutoring, 128
interoperability, 171

Java, 138
Javascript, 74, 115, 148
JSMath, 115

Khan Academy, 145, 147–148, 160
Knewton, 160

Lambert W function, 53
LaTeX, 69, 105, 114, 116, 153, 162
learning, 23–25, 45, 99, 128
  cycle, 33
Leonardo da Vinci, 26
linear algebra, 42, 48, 49
  matrices, 73, 85, 89, 111, 112, 115, 118
LISP, 79
logarithms, 1, 59, 73, 78
logic, 48, 100, 167, 169
  Boolean, 74
LOGO, 128

Maple, 61–63, 74, 76, 78, 102, 150, 153
MapleTA, 153, 160
Mathedge, 150
Mathematica, 61, 62, 74, 153, 169
Mathematical Association of America, 151
mathematical practice, 25
MathJax, 105
MathML, 67, 105
MathObjets, 153
Mathwise, 74, 131, 132, 148–150
MathXpert, 68, 100, 154–159, 162, 163, 166, 168
matrices, 89, *see* linear algebra, matrices
MatTaFi, 117
Maxima, 8, 61, 62, 64, 74, 76, 83, 98, 100, 105
mediant, 98
Messiah, 26
model solution, *see* solution, worked
modelling, 20, 21, 26, 28, 48, 166

Moodle, 8, 10, 13, 105, 115, 139, 140
motivation, 21, 25, 30, 32, 149, 160
MSOR Network, 1, 103
multiple choice, 2–5, 98, 103, 111, 130, 139, 140, 148
multiple response, 3
multiple steps, 107–111, 114, 131, 132, 154
MyMath, 160
MyMathLab, 160

National Assessment Bank, 135
National Problem Library, 152
Newton, John, 55
notation, *see* syntax
noun, *see* procept
NUMBAS, 69, 74, 132
numbers
  complex, 93, 118, 145, 149, 154
  floating point, 89, 112, 122
  Gaussian integers, 93
  irrational, 75
  pseudo-random, 39
  random, 75
  rational, 11, 42, 78, 89

OpenMark, 130, 138–140
OpenMath, 67, 105
order, term, 81
Oughtred, William, 55, 56

Paciolo, Luca, 55
paradox, 100
partial credit, 11, *see* multiple steps
Pascal, 131
Pass-IT, 132–136
Perl, 153
pie-chart, 135
plagiarism, 32, 38
Plympton 322, 37
polynomial, 49, 53, 73, 89, 154, 163–164
  CAS representation, 78
  cubic, 42, 45, 49, 81, 85
  Gröbner basis, 81, 96
  quadratic, 31, 39, 40, 42, 44, 57, 91, 95, 109, 128, 147
potential response tree, 106, 112
practice, 41, 43
problem-solving, 25, 29, 54, 151, 170

procept, 64, 79–81, 91
production rules, 60, 88, 146, 155, 157
  buggy, 97–99, 113
profile, *see* reporting
programming, 129
  assessment of, 127, 137, 168
proof, 20, 26, 67–68
  assessment of, 24, 25, 47, 48, 166–169
  by induction, 163, 166, 168
  process of discovery, 20
punched cards, 127

quadratic, *see* polynomial, quadratic
question space, 39, 41
  definition of, 41
  mathematical, 42

ranges of permissable change, 41
real analysis, *see* analysis, real
recurrence relation, 39
reflection, 21
reporting, 10, 35, 99, 104, 113, 138, 151
Reverse Polish Notation, 62

SAil-M, 170
SCHOLAR, 132
science
  assessment of, 140

scientific units, 19, 28, 57, 95, 97, 151
Scottish Qualifications Authority, 135
seed, *see* numbers, pseudo-random
semantics, 73
series
  harmonic, 40
sets, 79, 89, 112, 153
simplification, 75, 79, 86–88, 140
solution, worked, 13, 14, 34, 40–42, 74, 105, 106, 154
STACK, 4, 9–18, 74, 75, 95, 102–126, 140, 151, 152, 155, 163, 166
  development history, 8, 102–103
standards, 113, 123, 170–172
statistics, 20
steps, *see* multiple steps
surds, 76, 84–85, 89, 155
syntax, 5, 11, 53–73, 79, 80, 82, 102, 105, 110, 111, 130, 131, 151, 168
  ambiguity, 60, 80
  case sensitivity, 90
  cultural differences, 58
  expression trees, 79, 83, 91
  informal, 11, 66, 112
  interval notation, 151
  students' difficulties, 66–67, 123–124, 151, 160, 168
  units, *see* scientific units

T-algebra, 159
taxicab geometry, 44
Teaching and Learning Technology Programme, 148
technology
  student use of, 26, 147
TEX, *see* LATEX
Theorema, 169
tincture, cephalic, 23
Toolbook, 150
tree, *see* syntax, expression trees
trigonometry, 73, 87, 88, 142, 154, 165
TtH, 114
tutoring, 157

unary minus, 62, 69, 80, 90–91
undecidability, 85, 163
unification, 90
unit testing, 113
units, *see* scientific units
user profile, *see* reporting

variety, algebraic, 96
verb, *see* procept
Viete, François, 55

WeBWork, 70, 75, 150–153
Weierstrass substitution, 87
word problems, 27–32, 95